MODERN
ENGINEERING PRACTICE

MODERN ENGINEERING PRACTICE

Ethical, Professional, and Legal Aspects

D. ALLAN FIRMAGE, P.E.

Professor of Civil Engineering
Brigham Young University
Provo, Utah

Garland STPM Press
New York & London

15 14 13 12 11 10 9 8 7 6 5 4 3 2 1

Library of Congress Cataloging in Publication Data

___Firmage, David Allan.
  Modern engineering practice.

  Bibliography: p.
  Includes index.
  1. Engineering ethics. 2. Engineering—Contracts
and specifications. 3. Engineering law. I. Title.
TA157.F52   174′.9′62   79-23450
ISBN 0-8240-7108-5

Published by Garland STPM Press
136 Madison Avenue, New York, New York 10016

Printed in the United States of America

Dedicated to my wife
Margaret
who gave me encouragement and
support in the day-to-day writing
of this book.

CONTENTS

PREFACE

Every responsible person understands the need for ethics and integrity in all social and business relationships. Engineers and architects often become involved in complex projects costing large amounts of money. Under such situations, pressures and temptations of questionable conduct can be great. Design and construction professionals should have training in ethical conduct for all facets of their engineering and architectural activity.

This book is written with the purpose of training the student or practicing engineer/architect in standards of ethical conduct, responsibility as a professional, and the principles of engineering contracts. Since a valid contract of any kind is a legally enforceable agreement, the engineer/architect should have a basic understanding of all aspects of contracts with which he may be involved. No attempt is made here to train the student or professional as a lawyer, but a good understanding of the text material will help prevent legal pitfalls.

The material in this book is suitable for a two-to-three-credit semester course. In a three-credit semester course, the instructor may want to supplement the text material with case histories and other references fitting to the specific course syllabus. The references at the end of the chapters contain information on special topics. A course tailored to the contents of this book would be in keeping with the suggestion of the Engineering Council on Professional Development (ECPD) (changed to AAES in 1980) that all engineering programs have a required course covering the topic of engineering ethics.

The presentation of the subject of ethics in this book has not been confined to the first few chapters; rather, ethical considerations have been included in all chapters. Contract law and contracts should have an ethical base, and action choices should be governed by equity, honesty, and integrity. The subject of engineering contracts should not be taught without ethics as a prime consideration.

It was necessary to use the third person personal pronoun in many places in the book. The author is well aware that both males and females occupy engineering and architectural positions. However, to avoid the boredom of reading "his-her" wherever the personal pronoun is necessary, the masculine form only is used.

This book is written primarily for the purpose of training professional people in the subject area. It is not for the purpose of rendering legal advice or professional services. The author and publisher assume no responsibility for any action taken by a reader of this book.

The author expresses many thanks to Diane, Susan, Karen, Lynae, Lyona, and Joyce, who each at one time or another worked in the typing of drafts or the manuscript. Thanks also go to several engineers who reviewed the manuscript and made suggestions.

It is hoped that the learning experience gained from a study of this book will prepare the young engineer/architect to cope with the real world of his chosen profession with a better understanding of ethics, business relations, and contracts.

MODERN
ENGINEERING PRACTICE

1. ETHICS—ITS PLACE
IN SOCIETY

Ethics is defined as a set of standards of right and wrong; or that part of philosophy dealing with moral conduct, duty, and judgment. It can also be classed as the study of right and wrong. Webster's unabridged *New International Dictionary* [1] copyrighted in 1934, has a long definition of ethics and the classification of ethical theories. One of the principal ethical theories is the consideration, "happiness to be the greatest good." A second theory is that of perfectionism or self-realization; and the third theory is based upon man's relation to the universe or to divine law. All three theories come into play in a discussion of engineering ethics or the ethical standards of the engineer or even, in broad terms, the ethics of a professional person.

As a background to a study of ethics, the three theories of ethics just mentioned will be investigated further. With regard to the theory that happiness is to be the greatest good, some religious beliefs center on the theme, "man is that he might have joy." Many great teachers from Aristotle (350 B.C.) to the present have declared that to have joy, man must live by rules of conduct which consider others first and self second. One could discuss the meaning of joy and what is true joy and if it differs from happiness. However, the joy that is considered herein is that of long-lasting happiness that is brought about by a feeling of well-being. History can verify that societies that had high standards of conduct wherein people's actions were such that there was great concern for people other than oneself were societies of a high level of happiness. It is also without argument that where people lived with no concern or regard for others great misery and chaos prevailed. A classic story [2] of this lack of moral conduct is the history of the Confederate Civil War prison camp, Andersonville.

1

The second theory of ethics is perfection of the individual. This is based upon the idea that man can be classified as to the elevation of the personality (or soul) and that man himself will benefit (in personal well-being and inner peace) by rising to the highest level possible. A high level of perfection can only be achieved through high performance in obeying ethical standards. This theory may appear to be self-serving, but in serving oneself, one has to be mainly motivated to the well-being of others since ethics is the behavior of people as they affect others.

The third theory of ethical behavior is related to obeying divine law. Such is the foundation of all religious beliefs that have survived the test of time. Standards of living can be considered to have first come from God. It can go without argument that God's moral codes are for the purpose of making life more enjoyable and for bringing the greatest good to His people. Divine law is written in general terms and is nonspecific with regards to day-to-day relationships of people involved in the practice of professions. Man himself must then take part in setting standards. The influence of religious movements upon the ethical behavior of man cannot be taken lightly. More discussion of this will be presented later on in this chapter.

In a changing society, which most parts of the world have experienced in the last hundred years, there is the question of absolute or relative ethics. Certainly there are some absolute truths of moral behavior. Taking of life and stealing have been moral taboos in most societies for many hundreds of years. However, one may ask, "Is the taking of life and stealing never permissable?" Many would say that the taking of life is permissable when used for defense of self, others, or country. Who would blame a starving man for stealing food? Although most people agree that there are exceptions to any moral law, there are others who believe that there are no exceptions. This is the problem in formulating a code of ethics or enforcing one within a profession. However, to abandon an endeavor to prescribe a code of ethics because not everyone will agree upon its contents, wording, or enforcement is to abandon society. Man knows by history that the survival of oneself, family, and friends depends upon society accepting and abiding by a code of ethics.

Laws define a minimum of what man must do or must not do. They have to be finite in scope and understandable even to the unlearned. Ethical standards have a higher dimension than law in also directing what one "ought to do."

Standards of acceptable behavior change with time and may not be the same in different societies. One who has traveled in different regions of the world knows this to be true. Past history has shown that society in general tends to become more liberal. This is more so in those forms of conduct that more or less affect the individual; but there has been less acceptance of lower standards in those actions that affect others. Some actions that are accepted in a culture even though unlawful may be very unacceptable in other cultures. The acceptance of so-called "under-the-table-payments" to government servants may be a way of life in some countries, but in the United States it caused the resignation of a vice-president of the country. However, in countries where payments or bribes to officials are condoned, doing business is more difficult and distasteful, and the economic development of the country is retarded. Although a few individuals may temporarily benefit, society as a whole suffers.

The early Greek sophists believed that good or evil was determined by convention and not by nature. In essence they believed that man judged all things according to his own condition and the way things affect him.

Plato and Aristotle argued in the opposite vein. Plato believed that good is not a matter of opinion, but an object of knowledge.

In the *Republic* Socrates advised "to learn and discern between good and evil."

Locke and Kant believed in the scientific character of ethics, but Locke understood man when he said, "the desire of esteem, riches or power makes men espouse the well-endowed opinions in fashion."

Kant taught that the two major parts of philosophy are physics and ethics, and both are on equal footing.

It is evident that many of the "thinkers" believed in the absolutism of right and wrong. However, there were others, notably Spinoza, who wrote, "Good and evil indicate nothing positive in things considered in themselves, nor are they anything else than modes of thought." He defined "good" as "that which we certainly know is useful to us." He believed in man being controlled by law and he taught that only when men live together in a civil society under the law can it be decided by universal consent what is good and what is evil.

It is not the author's intention in this book to treat all the philosophical aspects of good and evil or to try to resolve the question of whether good is absolute or relative to the fashions of the day. This argument appears to take the background to Kant's statement of great worth, "Morality is not properly the doctrine of how we should make ourselves happy, but how we should become worthy of happiness—by doing our duty."

Ethics is related to human rights. Human rights can be considered in democratic countries as, "life, liberty, and the pursuit of happiness." It can also be defined as the absence of fear. It would be difficult to argue that the absence of fear is not a worthy goal of any society. Ethical behavior enhances the absence of fear or concern for one's rights when dealing with others. Everyone should have the right to expect honest, forthright behavior from those with whom they have relationships whether on a strictly social basis or in a business activity. If society expects or desires such behavior from others, then each individual should accept the responsibility of individual ethical behavior.

There is a weakness in human nature that leads to accepting the attitude that "when in Rome do as the Romans", and believing that all others in the business community are willing to take advantage of the weak, foolish, or less astute. This type of behavior is not the norm in the area of technical activity. A few bad examples can give an erroneous impression to the uninformed. Ethical conduct in the business of engineering/architecture and construction is the way, and each individual entering this activity should follow suit with a sense of strengthening the ethical standards of his profession.

An important aspect of morals and ethics is how these relate to the manner in which people live together in a relative degree of harmony. The past history of man shows a trend toward government by the people and for the people. Great steps in this progress were the Magna Carta in England and the formation of the Constitution in the United States. The principle of self-government was looked upon as utopian in many parts of the world. Although this form of government has flourished in America and several other countries, it has had limited success elsewhere. Self-government has sputtered and been abandoned for totalitarian rule in several countries. This pattern has been typical in a number of countries which have rejected colonial rule since World War II. Much, if not all, of this failure can be attributed to the lack of self-discipline. Self-discipline is voluntary adherence to a set of standards that places the good of others above our own selfish desires. Self-government without self-discipline is self-defeating. This truism is applicable to national governments as well as for local levels and beyond to professional groups of any kind.

At a commencement address at Brigham Young University in 1979, Gordon B. Hinckley said, "Integrity is the heart of commerce in the world. Every bank president knows that even with all the regulation and with all possible safeguards, in the last analysis the strength and safety of the institution lies in the integrity of its people. As with banks, so also with merchants, politicians, professional men and women. Without personal integrity, there can be no confidence. Without confidence, there can be no prospect of permanent success."

Through his experience man has developed a knowledge of norms that can bring to pass goodness and thus, greater happiness. This is an approach to ethical development that is a compliment to moral standards which have been received through religious channels.

It would be unacceptable to thinking people that man could survive without laws. It is also unacceptable to those in the professions to see no need for a code of ethics that would not only be a guide for those operating within the profession, but also a restraint to unethical behavior to those practicing the profession. Ethical principles are only of theoretical worth unless they are known, understood, and used in the practice of a profession. To be of value to society, ethical standards must be voluntarily accepted and enforced by those within the profession. Being neither for nor against moral rules is a form of indifference that can lead to a weakening of society. Dante stated, "The hottest places in hell are reserved for those who, in a period of moral crisis, maintain their neutrality."

Whether one believes in a heaven or hell, it is not hard to understand and believe that life on this earth, within or outside a profession, will be much better when society governs its conduct in conformance with high moral standards, and to also believe that such standards can in general be defined. A professional who has a concern for serving his fellow man and who wants to operate in an environment of well-being and good feeling toward his fellow professionals, will recognize the worth of an ethical code of conduct and hopefully be willing to live the standards as set forth by the code. Adherence to ethical principles should be in a spirit of the law and not just the letter of the law. William Shakespeare put it in words that have become immortal, "To thine own self be true, and it must follow, as the night the day, thou canst not then be false to any man."

There are records of ancient codes that were given to people of early periods. As soon as men started living in close proximity to one another, whether in 5000 B.C. or A.D. 1900, it became necessary to formulate some rules that would govern their behavior. Such rules, when obeyed, are intended to reduce friction with one another and to make life more pleasant for people as a whole. These rules were to some extent restrictive on the individual, but were for the betterment of society as a whole.

Restrictions were intended to prevent the actions of one party from infringing upon the well-being of another party.

Wise people who had a sense of right and wrong and who had some ruling or religious responsibility were the ones who wrote these ancient codes or laws. Undoubtedly, there were many written laws or codes covering man's behavior in past history. Some rules have survived to the present, while others have been lost in the ruins of the past. We do have records of those codes that have had the most influence on civilization and are considered by people of the present day as standards to govern and shape one's life. Those that have survived and are followed by people at this time all have a religious background or orientation. Some of these ancient codes are briefly described in the remainder of this chapter.

The Ten Commandments. These ten rules of behavior are part of the creed of the Judeo-Christian peoples. The belief is that Moses received these commandments from God on the Mount of Sinai (Exodus 20:3-17). These were laws for the Children of Israel who had been led from captivity in Egypt. Since the Israelites had been corrupted by living in idol-worshipping Eygpt, it was necessary to have a formal code that would govern their actions and thoughts. These Ten Commandments are accepted by those of the Jewish and Christian faiths as having come from God through his prophet Moses.

The Ten Commandments are dated by historians as having originated about the 14th century B.C. These commandments have been the foundation of common law in many countries throughout Western civilizations. The commandments of "Thou shalt not steal" and "Thou shalt not kill" have become the written law of most countries. Several of the commandments are strictly religious and the two pertaining to adultery and covetousness are to control the lusts of man. Whatever one's religious beliefs, it has been accepted by mankind that as rules of living the Ten Commandments benefit the well-being of the human race.

Both the Old and New Testament of the Bible contain many scriptures that give guidance with respect to the relationships of one person to another. Exodus, Leviticus, Numbers, and Deuteronomy contain laws and rules that have been the basis of English common law. Such rules have shaped the ethical behavior of many people and have become a foundation for law and order. These religious writings have been responsible for the orderly process of government in several areas of the world. The thought that man could live in harmony because he had high ethical and moral standards within his own character was the basis for the founding of democratic government. Such standards were and are necessary for the functioning of government wherein the laws and government operations are based upon individual freedom. Civil strife and instability of governments are a direct result of people within those countries failing to live by these highly ethical standards. No government can succeed and bring about real happiness for its people unless the people themselves are willing to live moral codes of conduct.

The Ten Commandments became the foundation of moral codes by the Jewish people and also all those people who have embraced Christianity. In addition, those of the Christian faith have believed and, to a greater or lesser extent, have practiced the teachings of Christ that pertain to the relationship of one to another. Such teachings emphasize the consideration for the well-being of others and a stress upon the spirit and not just the letter of the law. Since Christ lived and taught in a region and during a time of very little construction activity, the "Gospels" report very little concerning standards of conduct of a builder in the performance of his profession. This was not the case in Babylon about 2000 B.C.

The Code of Hammurabi. Hammurabi was a king of Babylon about 2000 B.C. History says that he was a successful and wise king. The standard for the rating of a wise king is based upon the code of laws attributed to him. These laws were written on a stone stela and found at Susa in Iran in 1901. During the reign of Hammurabi, Babylon was involved in much building and so the code also applied to the builder. These rules were very harsh for the builder who did not perform; they were also very definite, as shown by a translation of a portion of Hammurabi's Code:

> If a builder has built a house for a man and has not made his work sound, and the house which he has built has fallen down and so caused the death of the householder, that builder shall be put to death. If it causes the death of the householder's son, they shall put that builder's son to death. If it causes the death of the householder's slave, he shall give slave for slave to the householder. If it destroys property he shall replace anything it has destroyed; and because he has not made sound the house which he has built and it has fallen down, he shall rebuild the house which has fallen down from his own property. If a builder has built a house for a man and does not make his work perfect and the wall bulges, that builder shall put that wall into sound condition at his own cost.

All major religions teach, and have taught, moral standards that pertain to the relationships among people. These generally stress honesty, kindness, and concern for other human beings. In broad terms they can be applied to the ethics of professional people.

A sermon attributed to Buddha contains the passage, "if he be covetous in his desires, fierce in his longings, malevolent of heart or mind, corrupt, careless and unrestrained, not quieted, but scatterbrained, and uncontrolled in his sense, that one is far from me and I am far from him. . . . Brethren, do ye live perfect in virtue, do ye live perfect in performance of the obligations . . . perfect in the practice of right behavior; seeing danger in the lightest faults. . . ."

Other Religious Ethics. Buddha founded a religion based on moral values. He taught that universal brotherhood born from the love of all beings is the basis of moral principles. A teaching principle of Buddha was that there can be no love greater than the love of man and no service greater than the service of man. He therefore preached a religion full of love for all living things. Buddha also taught that there were Right Views, which came from the knowledge of the four truths. One of these truths was Right Speech. Right Speech requires one to abstain from lying, slander, abuse, harsh words, and idle talk. Right Effort consists in conscious effort of improving one's moral condition. It is the eradication of evil. Buddha asked his people to live a good and moral life. His emphasis in religious teaching was greater for the living of a good life than in the belief of a divine being.

Tibetan religious proverbs state: "Whatever is not pleasing to yourself, do not that unto others." Confucius is attributed to having said, when asked if there is any single saying that one can act upon all day and every single day, "Never do to others what you would not like them to do to you."

Confucius believed in a society composed of an intellectual upper class, but unless this was a moral upper class, it would fail and degradation would take place. History has frequently proven Confucius to be right. Unquestionable examples have occurred in the 20th century.

Mohammed preached the necessity of honesty in all relationships between men. In Sura LXXXIII Mohammed said, "In the name of Allah, the Beneficient, the Merciful! Woe unto the defrauders: Those who when they take the measure from mankind demand it full, but if they measure unto them or weigh for them, they cause them loss. Do you not consider that they will be raised again unto an awful day, the day when mankind stands before the Lord of the Worlds."

The Law of Karma in the Hindu Religion says that one's lot in future existences is dependent upon the ethical acts in this life: "Those who are of stinking conduct here—the prospect is, indeed, that they will enter the womb of a dog, or the womb of a swine. . . ." The above doctrine is from the *Upanishads*. Though specific unethical acts are not as well defined in Hinduism as in some of the other religions, mistreatment of others is certainly condemned.

In the early 19th century a noted Shinto scholar of Japan repudiated the suggestion that because the Japanese had no native system of ethics, they must borrow such a system from Confucianism. He stated that only a depraved people need a code of ethics. If a people live naturally upright they have no need of a code.

The Shinto scholar may be right but since man most everywhere has not reached that elevated state ethical guidelines appear to be necessary.

No doubt there were teachers of moral standards long before Hammurabi and Moses, and we hope there will also be teachers of honesty, justice, and good will toward men up to the end of days. Ethical teachings will harmonize with religious values no matter what one's special beliefs in the divine. Even though one may espouse no religious creed, one can readily accept the principles that codes of conduct accepted and lived by the majority will in most societies result in a much finer moral environment than would otherwise be attainable.

REFERENCES

1. *Webster's New International Dictionary of the English Language.* 2nd ed. (Unabridged), Vol. 1. Springfield, Mass.: G. & C. Merriam Co., 1934.
2. McKinlay Kantor, *Andersonville.* N.Y.: Thomas Y. Crowell, 1955.

2. ETHICS AND PROFESSIONALISM

A. DEFINITION

People involved and working in a specialized activity which requires special skill, knowledge and mental concentration define this activity as a profession and themselves as professionals. Considering oneself a professional can satisfy vanity and ego or it can be a motivating element in the striving for excellence and service.

The definition of a profession can best be given by the requirements and attributes of professional practice. However, the American Society of Civil Engineers (ASCE), in the printed *Official Register* defines a profession as "The pursuit of a learned art in the spirit of public service." It expands the definition with the following:

> A profession is a calling in which special knowledge and skill are used in a distinctly intellectual plane in the service of mankind, in which the successful expression of creative ability and application of professional knowledge are the primary rewards. There is implied the application of the highest standards of excellence in the educational fields prerequisite to the calling, in the performance of services, and in the ethical conduct of its members. Also implied is the conscious recognition of the profession's obligation to society to advance its standards and to prescribe the conduct of its members.

The Engineers Council for Professional Development (ECPD),* a council with representation from the major engineering societies, has given the attributes of a profession as follows:

1. It must satisfy an indispensable and beneficial social need.
2. Its work must require the exercise of discretion and judgment and not be subject to standardization.

*Reorganized as American Association of Engineering Societies (AAES) in January, 1980.

3. It is a type of activity conducted upon a high intellectual plane.
   a. Its knowledge and skills are not common possessions of the general public; they are the results of tested research and experience and are acquired through a special discipline of education and practice.
   b. Engineering requires a body of distinctive knowledge (science) and art (skill).
4. It must have group consciousness for the promotion of technical knowledge and professional ideals and for rendering social services.
5. It should have legal status and must require well formulated standards of admission.

ECPD has also stated what one who claims to practice a profession must do:

1. They must have a service motive, sharing their advances in knowledge, guarding their professional integrity and ideals, and rendering gratuitous public service in addition to that engaged by clients.
2. They must recognize their obligations to society and to other practitioners by living up to established and accepted codes of conduct.
3. They must assume relations of confidence and accept individual responsibility.
4. They should be members of professional groups and they should carry their part of the responsibility of advancing professional knowledge, ideals, and practice.

Although many working groups may classify their activity as a profession, it does not merit such classification unless the activity is conducted on a high intellectual plane and, as such, requires special training and knowledge in which this training and knowledge is applied with care and judgment. A profession must, as a group, promote the dissemination of knowledge learned through the practice of the profession. True professions will be involved in the education process, not only in providing information to one seeking entrance into the profession, but also to the continuing education of persons already within the profession.

Activity to restrain those who are seeking entrance into the profession by legitimate means is not in keeping with a professional. William H. Wisely, emeritus Executive Director of the American Society of Civil Engineers makes a salient point of professionalism when he states [1], "The obligation to give primacy to the public interest is the very essence of professionalism. Without this commitment, the effort of a group to seek elite status as exponents of a body of specialized knowledge is but a shallow and selfish charade, no matter how sophisticated that body of knowledge may be or how rigorously it may be pursued."

Voluntary groups within our society have sets of "behavior rules" that are formulated to provide a pleasant environment within the association and prevent encroachment upon the well-being of others. Very often the rules also provide protection from outsiders. A most marked example of this defense from encroachment from the outside is the labor unions.

A professional who is especially qualified in a particular field is expected to perform according to a standard of care that would be expected of any other similar professional and under the same circumstances. This has been the basis for judgment in most liability cases involving professionals. If the objective is service of the highest quality and there is adequate training and experience, then the professional should not be found lacking in his performance. Superior performance is a result of adequate preparation and knowing that what one is doing is right and honest. "Coverup" and dishonesty stems from poor performance which is a result of inadequate preparation, inadequate execution, lack of effort, or a combination of all three.

Deviant behavior in today's complex society can be highly disastrous. Nonthorough engineering can result in failure of an engineering work that as a minimum can waste thousands of dollars and as a maximum take human life. David Novick [2] has stated some of the demands felt by a professional engineer when he says, "he must have a good working knowledge of business, finance, and law because survival in today's marketplace is no longer a matter of just common sense."

There are always groups within society that have a desire to tear down other groups which have achieved success in their activities and thus have reaped monetary rewards as well as public recognition. In recent times such action has become more prevalent as directed toward professional groups. Engineers have been accused of destroying the environment, lawyers have been blamed for unethical practices and having little concern for the safety of society. The medical profession has also been subjected to public limelight, being accused of primary concern for financial rewards and not the welfare of the patient.

All professions are moral enterprises that involve concerns beyond the application of technical principles. How well the professionals meet these moral obligations will determine the freedom of the individual professional enterprise. Already the medical profession has lost a great

deal of freedom of operation in some European countries and this wind
is already blowing in the direction of North America. If the professions
do not regulate the moral actions of their members then others will. And
as was stated in a newspaper editorial, "and it's down that road that
nations go from regulations to regimentation to tyranny."

A viable society must be aware of any shortcomings of its people, but
it must also investigate alleged wrongdoings honestly and fairly. In the
present day of rapid and total communication via TV, radio, and the
printed page, misleading information and wrongful or half-true statements can perpetrate grave injustices. The public cannot be indifferent
to the well-being of a profession, whether it be law, medicine, or engineering. A healthy profession is a plus to the health of society.

The engineering profession is not void of those who desire to "make a
fast buck," but the laws of nature are hard task masters and stern judges
of incompetent behavior. Engineering is considered one of society's
activities that has the highest of ethical standards. Personal opinion
polls have been conducted in recent years to obtain the public's feelings
toward various professional groups. In such polls engineering has rated
near the top in public esteem and judgment of ethical standards. In a
1977 Gallup Poll, engineers were rated in the top three, along with
clergymen and medical doctors regarding the question of ethics in each
profession. In the "very high" category clergymen had 62 percent, doctors 51 percent, and engineers 46 percent. In ratings of low or very low
opinion were clergy 6 percent, doctors 10 percent, and engineers 5 percent. The rating for engineers was the lowest of any category.

A professional is one who has to make decisions, possibly many in the
course of a day. Those decisions are based upon a set of data that has
been gathered. These data may be sketchy, or very extensive and complete, depending upon the circumstances of the project. The professional must apply judgment to these data and then make the necessary
decisions. The correctness of this judgment will be the result of the
individual's experience and also the experience of others as obtained
from the literature.

A professional engineer should be conscious of social problems in his
community and should be willing to freely devote a portion of his time
to the solution of these problems. The professional engineer is highly
qualified to assist in the solution of community problems, but heavy
business demands many times serve as an "out" for not being involved.
Many social and community problems are attacked by groups from a
purely emotional bias. What is needed in most cases is a careful assem-

bling of facts and data and then a comparison of all alternative solutions based upon the facts available. In this action procedure, the engineer is eminently qualified. The high progress of social development in the United States has been the result of people willing to give their time and talents in the service of their community, whether that community is political, professional, religious, or social. Involvement in community programs will enhance the skill of the engineer. The "rubbing of shoulders" with people from other disciplines as well as other walks of life will give him better insight into the nontechnical aspects of his engineering projects and the social responsibilities of engineers and the engineering profession.

One can think of ethics and morals in a limited sphere of the profession only, but in a broader scope technology has brought about political, economic and educational systems that are the result of technological progress. For such institutions to receive the full benefit of such progress they must operate under a system of ethical standards.

Henry J. Taylor, noted news analyst, once said: "Essentially the problem (of society) is one of integrity. In a home, in a business, in a nation, integrity is what upholds all. It is the weakening of integrity that seems to me to be the greatest illness everywhere. The grand corruption of our age in fact is the inability of so many eminent human beings the world over to practice simple honesty and speak the simple truth."

B. PROFESSIONAL SOCIETIES

It is a fact of human behavior that people of similar occupations tend to mingle and associate themselves not only at work but also in outside social intercourse. There is some lessening of this cohesiveness in today's complex society than what was found in several past decades. Nevertheless, people tend to feel more at ease and trusting of those who earn their livelihood in the same or similar occupation. Class distinctions are less and subjects for conversation are easily found. In addition to social comfort there is a feeling of mutual protection by organizing into occupational groups. This feeling of a desire to belong to a group brought about the early trade guilds of Europe and was a catalyst along with other causes for the forming of trade unions.

In the middle ages groups of merchants and craftsmen formed guilds. These guilds were organized for protection, control of their occupation, and social relationships. There were merchant guilds which, in some cases, became rich and powerful. They set standards of quality and prices and benefited from lower prices due to quantity buying of goods.

The craft guilds set rules for training and membership. There were three groupings of membership: masters, journeymen, and apprentices. These titles still carry over into the trade unions of today. Guilds became protection societies and limited membership so as to keep monetary returns high. Bitter disputes arose between merchant and craft guilds, and the wage earners in the guilds formed unions in opposition to the larger industries. Labor unions are an offshoot of the guilds, but guilds as they existed in the Middle Ages are gone.

In the early years of professional engineering, engineers associated together to discuss engineering problems and to learn from one another. These mutual associations led to the forming of the first engineering society, the Smeaton Society, in the late 1700s in London. John Smeaton was a famous British engineer. The Smeaton Society grew into the Institution of Civil Engineers. It was established in England in 1818 and incorporated in 1828. The aims and function of this society were well described in its charter of 1828. Civil engineering was described in the charter as

> the art of directing the great sources or power in nature for the use and convenience of man, as the means of production and of traffic in states, both for external and internal trade, as applied in the construction of roads, bridges, aqueducts, canals, river navigation and docks for internal intercourse and exchange, and in the construction of ports, harbours, moles, breakwaters, and lighthouses, and in the art of navigation by artificial power for the purposes of commerce, and in the construction and adaptation of machinery, and in the drainage of cities and towns.

Civil engineering covered a broad spectrum of subject matter in that time period as it does today.

Early in the 19th century, mechanical engineering became a specialty and the Institution of Mechanical Engineers was founded in Birmingham, England in 1847. George Stephenson, the "father" of railroads, was the first president.

In America the first engineering society was the American Society of Civil Engineers, founded in New York City in 1852. The purpose of ASCE was the dissemination of technical information to promote high-quality engineering, and to protect the safety of the public. It also provided social activity among the members. Membership has always been restricted to persons of proven engineering qualifications, through education and/or experience. From a handful of members in New York City in 1852 ASCE has grown to over 75,000 members in 1979. Nearly 7,000 members are citizens of countries other than the United States.

Today in the United States there is, besides ASCE, the American Society of Mechanical Engineers, the Institute of Electrical and Electronic Engineers, the American Institute of Chemical Engineers, the American Institute of Mining, Metallurgical and Petroleum Engineers, as well as many other specialized technical societies of smaller size. Most countries with a body of practicing engineers have their own technical societies. There are some international societies as well, such as the International Association for Bridge and Structural Engineers, with headquarters in Zurich, Switzerland.

All major engineering societies have several activities that involve the membership and are their *raison d'être*. The major activities are:

1. publication of technical papers written by the membership;
2. holding of conventions in which technical and professional presentations are made;
3. providing advisory information to governmental legislative and administrative bodies;
4. promoting excellence in engineering education and safeguarding the training of engineers through the accreditation of engineering colleges and universities;
5. providing continuing education programs.

The major technical societies in the United States are large operations involving thousands of people in responsible roles, and operating with annual budgets of several million dollars. The societies are democratic organizations with elected officers and appointed committee members. These societies have many hard-working members who devote large amounts of volunteer time to the promoting of better engineering and to the maintaining of high standards of conduct and practice. The engineering societies' budgets mainly come from the membership dues and sales of publications which are primarily purchased by the membership and libraries. In the United States there is no government subsidy of engineering societies. Learned societies are not subject to federal and state income taxes on the income earned through the sale of publications.

C. ETHICAL DEVELOPMENT IN MEDICINE

The first records of a national scientific medical system are of Greek origin. Greek medicine in its pure form was established about 500 B.C. The Greeks were not the first to practice medicine of one form or another. It is known that the Egyptians, Indians, Persians, Minoans, and quite likely the Chinese had medicines and special people as dispensers of medicine who might be called "doctors" today. However, the Greeks near the end of the 7th century B.C. developed a whole science of medicine. They established medical schools at several locations.

Hippocrates was born about 460 B.C. on the island of Kos which is off the coast of Asia Minor. (Kos was the location of one of the early Greek medical schools.) He took up the practice of medicine and through legend and some known writings (some by Plato), he has become known as the "father of medicine." There is a group of about 100 Greek books on medicine called the *Hippocratic Collection* that have survived. It is known that Hippocrates did not write all of these books; in fact, it is not known whether he wrote any. The *Encyclopedia Britannica* states that no definite answer can be given to the question, "Which of these works is by Hippocrates?"

The *Hippocratic Collection* contains the now famous Hippocratic Oath. There is some question as to the authorship of the Oath. Most likely it was composed over a considerable period of time. The Oath has been given the title of Hippocratic Oath more to honor Hippocrates than because of his authorship. Many graduating medical students take the Oath, which includes rules for the relationship between a doctor and his patients. Some parts of the Oath have only a slight relationship to present-day medical practice but passages such as the following are as relevant today as in Hippocrates' day:

> I will follow that method of treatment, which according to my ability and judgment, I consider for the benefit of my patients, and abstain from whatever is deleterious and mischievous. . . .

> With purity and with holiness I will pass my life and practice my art. . . .

> Whenever, in connection with my professional practice, or not in connection with it, I may see or hear in the lives of men which ought not to be spoken abroad I will not divulge, as reckoning that all such should be kept secret.

> While I continue to keep this oath unviolated may it be granted to me
> to enjoy life and the practice of the art, respected by all men at all
> times, but should I trespass and violate this oath, may the reverse be
> my lot.

This is the earliest code of ethics of a profession that is known to exist.
It is, of course, broad in substance but sets a tone for the responsibility of
a learned profession.

Ethics has been a concern of the medical profession for centuries and
remains so today. Not every doctor is above reproach, nor is every lawyer,
engineer, or member of any profession. However, the world today has a
medical service at their disposal that was rarely dreamed of a few decades
ago. In great part this is due to the honest, dedicated work of all facets of
the medical profession.

The *New York Times* of February 20, 1978 reported that at that date
over one-half of the 116 medical schools in the United States had regular
programs in medical ethics. This shows the concern of the medical
profession for standards of ethics among doctors.

The medical profession, as is true of the legal and engineering professions at the present time, is seeking an answer to the best way that ethics
can be taught. Some are skeptical that moral integrity or ethics can be
taught to a person after they reach the age of entrance into medical
school. Others of the medical profession, and possibly the majority, are
of the opinion that although great emphasis should be placed on admit-
tance to medical school of students with a good background of moral
behavior, the teaching of ethics has a place in the training of medical
doctors.

Discussion of where and when to teach medical ethics is a present-day
activity. The *Journal of the American Medical Association (JAMA)* on
March 6, 1978 had an article [3] on the advocacy of teaching classical
ethics at the bedside. This paper does not suggest that the teaching of
ethics should be moved from the classroom to the bedside but that the
teaching of ethics should be carried beyond the classroom lecture by the
ethicist to clinical ethics by the clinician. In this paper, which demon-
strates a relatively new approach to teaching medical ethics, M. Siegler
states, "The problems of developing a curriculum in clinical ethics can
be surmounted by increasing the participation of clinicians while maintaining the valuable contributions of ethicist-philosophers in the education of medical students and physicians."

In this same 1978 issue of *JAMA*, an editorial [4] by E. D. Pellegrino,
M.D. reinforces the idea that ethics is present at the bedside and not just

a part of the business relationship. He states, "Ethics like any primary discipline that supports the physician's endeavors, can survive only if it is explicitly linked to the physician's central enterprise. That enterprise is the act of clinical decision—the choice among the many things that *can* be done and those that *ought* to be done for a given patient in a given clinical circumstance. This is the physician's essential out—that which makes medicine more than a summation of primary disciplines in the sciences and humanities. It must be a right decision—one that is good for this patient."

A response to these writings by F. W. McKee, M.D., in the January 5, 1979 *JAMA* was,

> If an individual at the time of entry to medical school does not already possess innate and well-ingrained principles of truth, honesty, compassion, common sense, and other related human virtues, it will be a rare circumstance indeed that such ethical attributes will be subsequently developed by either ethicist or clinician. The student is better served by contact with good and moral, rather than bad and amoral, preceptors. The dedicated student will survive the influence of an unethical senior, but if there is not firm character in either, the day will not be saved, regardless of the opportunity for structured curricular attention.

There is no doubt room for both philosophies in ethical training. The professional should have exposure at an early age to moral integrity and a belief in the obligation to do good. Then his professional training should include a reinforcement of his moral belief by training and experience in situation ethics. This is the present ideal of professional schools. Almost without exception, reference forms, as part of the application to professional schools, inquire as to the character of the applicant. At an early age all young people should be aware of this concern for moral character.

D. ETHICAL DEVELOPMENTS IN THE PROFESSION OF LAW

Ever since man has had laws there has been the practice of law and thus lawyers. In absolute forms of government, lawyers have usually been on the side of the state and were (and still are) concerned primarily with the conviction and punishment of the accused. Only in countries where governments recognize the rights of the accused to a fair trial has a group of defense lawyers developed. Regardless of the role of the lawyer and the requirements of the client for legal help, a system of government by law requires a body of lawyers who are ethical and who place the needs of the client foremost in their priority of concern. Without lawyers who can be trusted to operate in an honest, ethical manner, laws used as a means of controlling society can become a farce.

In 1908 the American Bar Association (ABA) adopted Canons of Professional Ethics. Amendments followed and in the late 1960s the ABA prepared a Code of Professional Responsibility and a Code of Judicial Conduct. It was approved and became effective for ABA members on January 1, 1970. This ABA Code is very extensive; however, space and need justify only a brief review in this book.

The Preamble to the ABA Codes gives the case for the need of a professional code. The introduction to this preamble states:

> The continued existence of a free and democratic society depends upon recognition of the concept that justice is based upon the rule of law grounded in respect for the dignity of the individual and his capacity through reason for enlightened self-government. Law so grounded makes justice possible, for only through such law does the dignity of the individual attain respect and protection. Without it, individual rights become subject to unrestrained power, respect for law is destroyed, and rational self-government is impossible.
>
> Lawyers, as guardians of the law, play a vital role in the preservation of society. The fulfillment of this role requires an understanding by lawyers of their relationship with and function in our legal system. A consequent obligation of lawyers is to maintain the highest standards of ethical conduct."

The ABA Code of Professional Responsibility consists of three interrelated parts: canons, ethical considerations, and disciplinary rules. They define the type of ethical conduct that is expected of lawyers and also nonprofessional employees. A lawyer is expected to be responsible for the conduct of his employees and associates in all actions of professional representation of the client. The engineering codes do not quite specifically define this responsibility of the professional engineer to his nonprofessional employees.

The canons are similar in nature to the canons of the engineering codes in that they state in general terms the expected conduct of the professional in relationship to the public, the legal system, and the legal profession. The ethical considerations represent the objectives of general conduct toward which every lawyer should strive. They are not specific as to action, while the disciplinary rules are. The disciplinary rules are mandatory in nature. They define the minimum level of conduct below which no lawyer can fall without being subject to disciplinary action.

As in the engineering codes, there are no prescribed penalties for violation of the disciplinary rules. The penalty for one found guilty of violating the rules is determined by the peer group of lawyers appointed by the state Bar Association. The severity of judgment is based on the character of the offense and extenuating circumstances.

Investigation and hearings of violations of the ABA Code differ from that of ASCE in that the ABA handles such cases on a local state level, whereas ASCE action is on a national level.

An example of the three parts of the ABA Code of Professional Responsibility are given as follows:

CANON 4: A LAWYER SHOULD PRESERVE THE CONFIDENCES AND SECRETS OF A CLIENT.

Ethical Consideration EC 4-5:

A lawyer should not use information acquired in the course of representation of a client to the disadvantage of the client and a lawyer should not use, except with the consent of his client after full disclosure, such information for his own purposes. Likewise, a lawyer should be diligent in his efforts to prevent the misuse of such information by his employees and associates. Care should be exercised by a lawyer to prevent the disclosure of the confidences and secrets of one client to another, and no employment should be accepted that might require such disclosure.

Disciplinary Rules DR4-101(c):

(c) A lawyer may reveal:
1) Confidences or secrets with the consent of the client or clients affected, but only after a full disclosure to them.
2) Confidences or secrets when permitted under Disciplinary Rules or required by law or court order.
3) The intention of his client to commit a crime and the information necessary to prevent the crime.
4) Confidences or secrets necessary to establish or collect his fee or to defend himself or his employees or associates against an accusation of wrongful conduct.

The above example, from Canon 4, shows the nature of the Canons, Ethical Considerations, and Disciplinary Rules.

In the Code of Professional Responsibility there are nine Canons. Each canon could almost apply to the engineering profession. Canons 1 through 3 and 8 pertain to the operation of the legal profession. Canons

4 through 7 spell out the responsibility of the lawyer to his client. Canon 9 relates to personal integrity of the lawyer and the need for establishing and maintaining public confidence in the legal profession. This latter responsibility is a hard matter to accomplish, since many of the people who are involved in legal confrontations, whether civil or criminal suits, are going to be judged the "losers." Many losers will and do judge the lawyers (including judges) as unfair or even as dishonest. Trying to maintain public confidence is not an easy task, yet it cannot be taken lightly. Engineers do not have such a difficult role despite attacks by environmental groups.

Law schools have been teaching their students about professional ethics for many years. The judgment of success will vary depending on who is doing the judging and will certainly be biased by personal experience. It would, however, be hard to refute the hypothesis that the legal profession and society as a whole is better because of such an endeavor at teaching professional ethics.

There are over 150 law schools in the United States with quite different traditions, student and faculty roots, and backgrounds. The accreditation of these law schools is by the American Bar Association. The license to practice law after graduation from a law school is granted by the state. The control of examinations and monitoring of the action of practicing attorneys is given to a board that is appointed by the governor. This operation is of course authorized by the respective state legislatures. Since state legislatures have substantial numbers of lawyers within their ranks, the legal profession has been more closely controlled by law and the state boards than the professions of medicine or engineering. One result is that legal education is controlled much more closely by legislatures than either medical or engineering education.

Courses in the area of ethics and legal responsibility have been offered by many, if not most, law schools for several years and recently such courses have become mandatory. The American Bar Association now includes in its standards of education the following requirement:

> (iii) [Law schools] shall provide and require for all student candidates for a professional degree, instructions in duties and responsibilities of the legal profession. Such required instruction need not be limited to any pedagogical method as long as history, goals, structure, and responsibilities of the legal profession and its members, including the ABA Code of Professional Responsibilities, is encouraged to involve members of the bench and bar in such instruction. [Standard 302—1977].

It is common for questions on legal ethics to be included in the state bar examinations. California has a separate examination on legal ethics that is part of the bar examination. A definite catalyst for the training and examination of law students in legal ethics was Watergate. Since most of those indicted were registered lawyers, there were many editorials written in the public press about the need for higher standards of ethical conduct in the legal profession. One respected national magazine devoted many pages of an issue to the problems of legal ethics and the need for better training and enforcement of the ABA Code of Ethics. It went so far as to call the law profession a "sick profession." There has, without question, been forward action undertaken to strengthen the enforcement of adherence to ethical practices by lawyers.

Members of the legal profession are in a unique situation in that they are constantly dealing with the law and in many cases defending lawbreakers, and in doing so they are torn between two loyalties—on the one hand, loyalty to the client and on the other, loyalty to their profession and its ethical code. The role of the adversary can, at times, be difficult and very demanding of ethical behavior. The withholding or suppression of information that may be damaging to a client may be the "proper procedure" from a standpoint of legal maneuver, but it can also introduce mental and operational habits in the lawyer that may not be conducive to accepted ethical behavior.

Unethical conduct by lawyers in most cases is in a way similar to that in engineering, since it is the result of inexperience, ignorance, or inadvertence. Competition for business is sometimes keen and pressures build up. Sometimes wrong behavior can be attributed to the practitioner's being naive and being "trapped" by inexperience. Nevertheless, most actions can be assessed to be basically honest or dishonest by any law school graduate. The same can be said of engineering and medicine.

E. ETHICS IN BUSINESS PROFESSIONS

Since men started trading with one another there have been proprietary conflicts. Since a large part of the business world deals with buying and selling, the natural desire is to sell at the highest price or buy at the lowest price. This naturally leads to "What can I do to get the highest price?" and "What can I do to get the lowest price?" For ages merchants and buyers have competed over prices and many have resorted to all kinds of tricks, gimmicks, or whatever one has to do to get the best of a deal. The question of ethical conduct was rarely considered. The only guide to ethical conduct came from religious teachings with regard to the sin of avarice and the preaching of ecclesiastical authorities that "What profit a man if he gains the whole world, yet loses his own soul."

In the Middle Ages there were some church-state attempts to control the injustice of an unregulated economy. Some of these were to give guidance to the establishment of prices on basic commodities but, as always, the setting of prices in a free market society was almost hopeless.

In the time of the Reformation there were new insights and direction into man's vocational practices and his relationship to God. Luther and then Calvin were teachers of the principle that the manner in which a person conducted his labors and used his earnings had a definite bearing on his relationship to God, now and in the hereafter.

Religious reformers have, through the ages, taught that the problems of earnings, property, and disposition of earnings and property have an impact upon the well-being of society. These teachings did exert some control on business, especially in localities where there was a close relationship between church and state. What the church taught was also the common law of the state, and in alleged injustices brought before the government authorities, religious principles were the basis for rendering judgments. However, as church and state became divided in the Western world, a degeneration of ethical standards in the control of business dealings took place. The merchant became less impeded by restrictions imposed by church authorities. His religious control was likely limited to the effect of the Sunday sermon or his pangs of conscience during his periods of "confession."

The Industrial Age then compounded the complexities of moral behavior in the business world. Relationships between employer and employee became very impersonal. As men, women, and children started working in factories, they became numbers and lost their identity as human beings. Moral injustice in the business community has been the subject of many popular writers such as Shakespeare in *The Merchant of Venice* and Dickens in *A Christmas Carol.*

The ills imposed upon workers by the Industrial Age brought about much social unrest. It developed class feelings based upon economic position. Turmoil was on the increase and as evil begets evil, there were crimes committed by both management and labor. There were many who advocated controls upon the business community by the state. There was some self-regulation by industries and not all management was controlled by the profit motive only.

In the late 19th and early 20th centuries spokesmen for the working class promoted the idea that the only way to eliminate the ills of industrialization was to turn the management of industrialization over to the state. Marx and later Lenin were the most noteworthy proponents of this concept. Lenin turned the philosophy of state control of means of production into a political revolution that was successful in Russia in 1918. Mao in China in 1949 and other revolutionaries in other countries of the world made capital of the fact that the masses of people had hoped for material improvement in their lives and rejected the economic system which had failed to better their lives. A failing in the thinking of the early and even present day revolutionaries is the fact that no system will be any more ethical than the ethics of the people who run the system. There is no reason to believe that a person who rises from the laboring class will be any more ethical than a person from management or royalty.

A lack of ethical standards in the business community has brought much hardship and turmoil to mankind. As has been stated by F. H. Knight [5], "Economics and ethics naturally come into relations with each other since both deal with the problem of value." A gross lack of moral standards in business affairs will ultimately bring about demands for redress from the populace and also demands upon the state to enact laws that will prevent business practices which are detrimental to the economic well-being of a substantial segment of the people. Time has shown that increasing dishonesty will ultimately result in the loss of freedom to the individual, either by incarceration or by restrictive laws. What may in the beginning appear to be liberty will, if misused, result in restrictions not only on the offender but upon whole classes of people. Even in the free society of the United States many laws have been enacted that control business practices. A large percentage of these were enacted during or since the Great Depression of the 1930s. Present political debate is increasingly occupied with subject matter related to control of business practices. At this writing, bills have been entered into Congress to provide a Consumer Protection Agency. There is an opposing opinion that the economic well-being of the country is being restricted by too many laws and too much government interference in business. Higher ethical standards would do much to thwart further laws and any more governmental control. A pointed example of this is a statement by a Federal Trade Commissioner who in 1979, when speaking at the annual convention of home builders, said, "I say to you today that as home-builders you have a choice: either you can independently decide to make self-regulation work or you can brace yourselves for a full-scale, hard-hitting regulation from the government. It's that simple."

The concept of *laissez faire* will only work in an environment of high ethical standards. A recent statement by a university sociology professor in a newspaper article on the law and regulatory provisions was, "In this country (U.S.A.) there is almost an expectation of law violation in certain areas of life, such as politics and business; thus the idea of legitimized crime."

Communism has propagandized the idea that it can and does eradicate corruption in government and business. This offer of freedom from graft and lawlessness has appealed to people of numerous countries, especially where they do not comprehend the loss of freedom offered as an alternative or where there already existed little freedom. Corruption in the Diem government and those that followed was a major factor in the fall of South Vietnam. At the time of this writing there is turmoil and bloodshed in other countries that could have been avoided if the political and business leaders had understood and accepted a moral code of conduct that considered the well-being of all people. Selfishness, greed, and avarice have certainly brought grief to the world.

There are business groups that have codes of conduct that are a guide to their respective members. It is not possible to list these here or discuss the code contents or means of enforcement. Let it suffice to say that people of Western culture operating in the business arena fully understand the white and black areas of right and wrong; however, the gray areas of business relations is where most questionable conduct occurs. This is the area of not breaking the written law, but still taking advantage of another person's weakness, innocence, or folly.

There are many who feel that there should be a considerable upgrading in the ethics of the business community if the free enterprise system is to survive. Some believe it will have to come through government regulation, while others would hope that various business groups would regulate themselves. Government regulation eliminates the free enterprise part of the system.

In November 1977, a nonprofit organization called the American Viewpoint announced the formation of the Ethics Resource Center for the purpose of supporting the proposition that "individual honesty and self-discipline are indispensable to our economic freedom." Also in 1977, Gerald L. Martin, Vice President of Land O'Lakes, Inc., of Minneapolis made an appeal in the September issue of *Business and Professional Ethics* to individuals in business to join him in a Society for

Business Ethics. Such a society would be for the purpose of developing a reasonable set of guidelines of personal ethical behavior.

A positive aspect of business ethics was presented by Robert C. Parker, Vice President of International Harvester Company and chairman of the Ethical Standards Committee of the National Association of Purchasing Management. In an interview with *Boardroom* (Vol. 8, No. 8, 1979), he stated, "The minute a buyer is perceived as being anything other than ethical, it starts costing the company money. Word travels fast among suppliers, even about subtle differences in ethical standards among different departments. Top managers can and should influence the signals that buyers transmit to suppliers." A suggestion Mr. Parker made was to frame and hang standards of purchasing behavior on the reception wall area where visiting salespeople can't fail to notice.

There have always been those "crying in the wilderness" for reform in business practices, but it is hoped that their calls will be heard in the coming years so that people everywhere can enjoy economic freedom in a spirit of brotherly concern.

F. ETHICS IN GOVERNMENT

Governments control large amounts of money. In some countries the government controls a large proportion of the fiscal activity of the country. Though this is not the case in the United States, the amount of money spent by all units of government is enormous. Some single projects will expend millions of dollars. The temptation to siphon off some of this money for private use is very great. A small minority of government employees are not able to resist this temptation. Engineers and architects are involved with units of government in a wide diversity of activities—from weapons of defense to dams and bridges. When dishonesty occurs it quite often involves contract professionals, as well as government employees. A considerable proportion of the professional conduct cases tried by the American Society of Civil Engineers are ethical problems involving engineers and government officials.

Professionals doing business with governmental units should take great care in making sure that all expenditures of funds and all fiscal transactions on government contracts are honestly and ethically used. Conforming to all government rules and regulations as well as all contract clauses should be the standard procedure. Any deviations can be considered unethical activity and could bring civil or criminal charges.

Government contracts have special regulations with different procedures for various units of government.

Government employees who work in a professional area and administer contracts with professional firms should handle these contracts fairly and honestly. Underhanded actions to make the work of the contractor difficult or giving a special favor to a contractor with or without favors in return are most reprehensible. Knowledge of such dishonest action should be quickly reported to the proper authorities. Gifts of any size or kind from a contractor should be refused. Placing oneself in a compromising position for having accepted small or large favors is demeaning and can lead to deep involvement and the possibility that it might be viewed as a criminal act. The best policy of a government employee is to refuse any offer of a favor from any organization doing business with the government. This can be done in a polite and direct manner when first approached. This will most likely be the end of any unethical overtures.

As was previously stated, most people working in or having business with government have a high degree of integrity. The great volumes of honest activity go unnoticed—it is the few cases of corruption that appear in the news media.

History shows that corrupt governments will eventually come to a downfall. Basically most people want and expect honest civil servants. A large measure of the greatness of the United States in freedom, high standards of living, and progressive environment is due to the minimum amount of government corruption. However, vigilance is always necessary. Professionals should bear a major responsibility in trying to set and maintain honest government. Each generation has that responsibility to the generation that will follow. Making the world right for those who will follow is the noblest of actions. Proper behavior patterns should be established early in a career. Deviations will then be less tempting.

In the last half of the 20th century, the environment of land, air, and water has become of major concern. Issues pitting the quality of the environment against construction projects have become very intense. New government regulatory agencies have been created that are staffed by many engineers and scientists. An important role that professionals play in these agencies is the gathering and evaluating of such data to provide assistance in making rational decisions. Laws are passed, projects approved and funded or rejected based in large measure on the information supplied by professionals. Therefore the collection and evaluation of such data or statistics should be performed with the greatest degree of care and integrity. Cases have been reported where personal preferences and biases resulted in the reporting of erroneous data or even the falsifying of information. Determining incorrect conclusions and then passing them on as fact is professional dishonesty and when such action is detected strong measures should be taken to replace such unethical personnel. Environmental problems are of grave concern and affect almost all activities of the people of a country. Unethical behavior, prompted by personal biases, should not be tolerated. The obtaining of the truth should be paramount in all evaluations.

G. THE GOVERNMENT'S ROLE IN PROFESSIONAL ETHICS

A major role of the federal government is the enforcement of laws passed by Congress. Over the years there have been numerous laws passed in an effort to control ethical conduct. Most of these laws pertain to business operations. The Sherman Antitrust Act is an example of such a law. Copyright and patent laws are also laws of ethical standards. The enforcement and interpretation of laws are the responsibility of the judicial branch of government. Sometimes interpretations of laws are not always the same as original intent.

The state governments play a major role in the operation of professions. Each state has a department of business regulation that has been authorized by the state legislatures. These boards are authorized primarily for the purpose of requiring competency for people to practice in a trade or profession. The state professional or trade groups have been instrumental in setting standards of qualification. These standards of qualification contain requirements of preprofessional experience, and the passing of examinations to prove competency.

Operating under the department of business regulations are state boards that are composed of people appointed by the governor, usually from a list submitted by a local chapter of a technical society. These boards review applicants for licensing and all preparation and grading of examinations. Recently, notably in California, state laws have been passed that require each professional and trade board to have lay people on the board. The rationale behind this is to prevent the boards from becoming too restrictive in the processing of applicants. It is very questionable that lay persons are effective on these boards and in a position to judge competency or the quality or fairness of examinations and their grading.

As stated, the regulation boards are primarily for the purpose of protecting the public from unqualified people operating in areas that require special training and expertise. The professional registration boards do some ethics monitoring, although they do not function as investigative bodies. The powers of the registration boards vary from state to state. Some boards can only revoke a license upon proof that the registrant practiced fraud or deceit in obtaining the certificate of registration or being found guilty of gross negligence, incompetence, or misconduct in the practice of the profession. The State of California Professional Engineers Act gives the rule for suspension and revocation of a license as follows:

> A board may suspend or revoke a license on the ground that the licensee has been convicted of a crime, if the crime is substantially related to the qualifications, functions, or duties of the business or profession for which the license was issued, or the ground of knowingly making a false statement of fact required to be revealed in an application for such a license. [Chapter 3, Section 490]

Such a reason as the above for revoking a license does not include any jeopardy for unethical practices. In the author's opinion the bribing of government officials in order to receive a design contract would be grounds for revoking an engineering license; however, the courts would have to decide if such were so for a particular case. In bribery scandals across the country, there has been great demand for the withdrawal of professional license. The result has varied from state to state and case to case.

State legislatures have been reluctant to write laws that would take away a person's means of livelihood and even more reluctant to give the power of such action to appointive boards. The state medical boards have found it very difficult to prosecute so-called "quacks" for practicing medicine without a license. Judges and juries are very reluctant to convict a person unless it can be proven that the unlicensed person actually caused damage. This is more difficult to determine in medical cases than in engineering practice, where a structural failure may have occurred.

In the 1979 session of the Utah legislature, a law was passed to strip the Utah State Bar of its power to license and discipline attorneys, and to transfer the responsibility to the state supreme court. The sponsor of the measure said there were complaints that the Bar Society sometimes functions as a protective society for attorneys. The law gives the high court authority to appoint committees that would license lawyers, discipline them when they commit minor infractions of rules laid down by the court, and revoke licenses of those found guilty of major violations.

One of the ironies of government legal activity is that the Federal Department of Justice has brought action against several professional societies because of clauses in their codes of ethics. One's first thought would be that the legal establishment should be attempting to enhance and build up ethical practices and not be spending their energies in demolishing codes of ethics. The following chapter will treat this subject in more detail.

It is not possible to control the ethical climate of any profession by means of laws passed by government entities. The possibilities of wrongdoing are too numerous and extensive for laws to cover all possible acts that may occur in the operation of a business or profession which may cause harm to the professionals or to society in general. Also, this means of achieving ethical behavior is at the expense of personal freedom. Furthermore, when man controls his own actions by a feeling of "it is the correct thing to do," there is an uplifting of the soul and the environment of human feeling is on a much higher plane than when action is due to external pressure. Instilling knowledge of "what is the correct thing to do" is the purpose of ethical teaching.

Because of the increasing awareness of the need to teach ethics in American institutions for higher education, the Carnegie Corporation awarded the Hastings Center (Institute of Society, Ethics, and Life Sciences) a substantial grant for a study on the teaching of ethics in institutions of higher education.

REFERENCES

1. W. H. Wisely, "Public Obligation and the Ethics System," American Society of Civil Engineers, Preprint No. 3415, October 1978.
2. D. Novick, "Requirements of Professional Practice," *Engineering Issues*, ASCE, April 1976.
3. M. Siegler, "A Legacy of Osler—Teaching Clinical Ethics at the Bedside," *Journal of the American Medical Association*, March 6, 1978, Vol. 239, No. 10, pp. 951–956.
4. E. D. Pellgrino, editorial, "Ethics and the Moment of Clinical Truth," *Journal of the American Medical Association*, March 6, 1978, Vol. 239, No. 10, pp. 960–961.
5. F. H. Knight, *The Ethics of Competition*. Chicago, Illinois: University of Chicago Press. 1935.

3. CODES OF ETHICS IN ENGINEERING SOCIETIES

A. INTRODUCTION

The engineer in his career will without question be confronted with circumstances that will require knowledge of what is right or wrong. How does the engineer acquire a knowledge of what is proper behavior? Standards of conduct do not come naturally; they have to be learned. Basic principles of honesty and integrity should be learned in the home during the early years of life. As the student progresses through the classroom, he will be presented with moral ideas to a greater or lesser degree. The student will also be influenced in forming sets of values by activities away from home and the classroom. These influences may be uplifting or degrading. By the time the future engineer reaches college, he will have been exposed to many thoughts and experiences that will shape his moral and ethical standards. As he studies mathematics and science and then applies them to engineering problems, he will learn that to be successful he must learn the truths of science and engineering. Engineering training teaches that there are truths and standards that must be learned and applied in order for engineering works to be successful. It should then be easy to believe that there also must be standards of ethical behavior that must be followed for the profession of engineering to operate successfully. Engineering requires mental discipline and it also requires moral discipline.

Ethical conduct in engineering is that action conforming to the law of the land as well as moral standards of the profession. The engineering societies have to a degree spelled out what the moral standards of the profession are—that is, beyond what the law requires. Engineering ethical codes are necessary in the present day technological culture, since the work of the engineer effects the lives of many if not all members of society, and most certainly all people of industrial countries. His work and the modus operandi of engineering is little if at all vaguely understood by most people. The public has to rely on the engineering profession itself to police its members and to see that quality engineering is performed and the business of engineering is conducted honestly and ethically. Integrity should be the motto of the engineering profession, a profession that is a people-serving activity.

The various engineering disciplines have taken upon themselves the obligation of providing standards of conduct which are called *Codes of Ethics.* Undoubtedly many new engineers enter into the profession each year who have not seen or heard of any rules of conduct applying to their chosen occupation. In recent years the engineering societies have been aware of this ignorance and an education campaign has been in progress. New engineers will quite likely be unknowledgable of ethical standards of the profession unless they have enrolled in a college course that teaches such or unless they join a professional society that has a Code. Unless such a Code is well publicized, the new engineer may still remain unfamiliar with all the nuances of behavior that he should and is obligated to follow. At the present time a course in engineering ethics is not a definite requirement for accreditation of a college engineering program by the Engineers Council for Professional Development.[1] However, it is certainly encouraged and such a course does meet one of the engineering category requirements. Neither the state Engineer-in-Training nor the Professional Engineer Examinations contain questions on engineering ethics. This is a subject for consideration by the engineering societies.

Whenever there has been wide publicity of unethical behavior on the part of people who handle or use public funds there is a clamor for new or more laws to prevent such future happenings. The law with its sanctions is a means for promoting ethical conduct, but the law alone will not provide for a pure climate of professional behavior. The law only provides minimum standards of conduct and as a last resort is a vehicle for punishment of the wrongdoer. The possibilities of unethical conduct are too numerous and extensive for laws to be written that will cover all the forms of conduct which may cause harm to the public and to the profession.

Rising consumer consciousness and actions by legislative bodies, government regulatory agencies, and the courts pose a threat to the traditional self-regulation practices of the profession. Unwelcome external restraints by government regulatory agencies have been imposed on professional societies and may again be imposed in the future. Self-regulation is better for society and will certainly result in a much more healthy profession which, as stated in a previous chapter, is also better for society.

The writing of codes of ethics is no easy task. Many people have struggled with the duty over the past years. The first American engineering code of ethics was adopted by the American Institute of Consulting Engineers in 1911 and was based on the code adopted by the Institution of Civil Engineers of Great Britian in 1910. The content was directed towards the relationship between consultants and clients. The AIEE adopted a Code of Principles of Professional Conduct in 1912. The first American Society of Civil Engineers' *Code of Ethics* was issued in 1914. Since the practice of engineering has changed considerably since these early dates, the codes have also changed.

There are diverse opinions with regard to the basic philosophy of engineering codes of ethics. W. H. Wisely, a former Executive Director of ASCE, has suggested [1] that a code of ethics of a profession has an ethic statement as the core that would stress the obligation to serve the public interest. Business etiquette would be relegated to a separate set of rules. He suggests that the unprofessional conduct of any engineer be judged by the professional ethic and not by the rules. Judgment would be by a group of peers. The rules would only encourage conformance. Wisely suggests the following statement as the professional ethic: *The engineer shall apply his special knowledge and skill at all times in the public interest with honesty, integrity, and honor.*

There are no doubt contrasting views to those of Wisely as to what a code of ethics should contain. In 1974, C. R. Schrader [2] said, "the interface between what we design and build and their effects on mankind are either not covered by existing ethical canons or leave doubt as to how the canons should be interpreted and implemented."

It is not possible to spell out the proper course of action which an engineer may take. Not all problems can be resolved with black or white answers; sometimes there are wide gray areas. The ethical course of action is studded with many pitfalls. Who "wins" by the decision; and is winning the game what really counts?

Recent thinking on ethical standards has become much broader, encompassing responsibilities in a greater area of activity than was the concern of earlier ethical codes. Many of the early codes, as mentioned, just related to the responsibilities of an engineer to his client. Later the responsibility of conduct affecting other engineers was added. Now, there is a growing feeling that ethical standards should also include the sensitivity of the engineer to the "correctness" of the entire project. Ethical problems of engineering practice now include social problems as well as technical problems. People feel that professional engineers now have responsibility for the social and environmental consequences of the projects for which they are responsible.

B. THE AMERICAN SOCIETY OF CIVIL ENGINEERS

The American Society of Civil Engineers has been very active in promoting a code of ethics and in enforcement of such code. The first code was adopted in 1914, and for many years ASCE has had a Code of Ethics containing thirteen articles. However, in 1971 ASCE, NSPE, and AIA were told by the United States Department of Justice that their codes of ethics were in violation of the Sherman Antitrust Act. The history of this action is given in Section E of this chapter. The Code of Ethics of the ASCE that had been in use for many years, and was the subject of the DOJ action is as follows:

It shall be considered unprofessional and inconsistent with honorable and dignified conduct and contrary to the public interest for any member of the American Society of Civil Engineers:

1. To act for his client or for his employer otherwise than as a faithful agent or trustee.
2. To accept remuneration for services rendered other than from his client or his employer.
3. To invite or submit priced proposals under conditions that constitute price competition for professional services.
4. To attempt to supplant another engineer in a particular engagement after definite steps have been taken toward his employment.
5. To attempt to injure, falsely or maliciously, the professional reputation, business, or employment position of another engineer.
6. To review the work of another engineer for the same client, except with the knowledge of such engineer, unless such engineer's engagement on the work which is subject to review has been terminated.
7. To advertise engineering services in self-laudatory language, or in any other manner derogatory to the dignity of the profession.

8. To use the advantages of a salaried position to compete unfairly with other engineers.
9. To exert undue influence or to offer, solicit or accept compensation for the purpose of affecting negotiations for an engineering engagement.
10. To act in any manner derogatory to the honor, integrity or dignity of the engineering profession.

In reading the above Code it is observed that the Code covered a number of responsibilities of civil engineers. The emphasis was, however, toward the business ethic and the action of engineers that would effect other engineers. Article 3 of this code was the target of the DOJ action. This article prohibited ASCE members from submitting priced proposals for engineering contracts when price is to be used as the principal criterion for selection of the engineering firm. This prohibition was interpreted by the DOJ as being in violation of the Sherman Antitrust Act.

In view of this confrontation (See Section E for details) ASCE rewrote their Code of Ethics, effective January 1, 1977. The new code is listed in the Appendix. It is noted that the new code is much more inclusive and in a different form from the old code. It also removed "business restraints" as contained in the old code.

The form of this 1977 code has changed considerably from the previous one. It sets forth the four fundamental principles. These same principles are accepted by the Engineers Council for Professional Development (ECPD—now the AAES). They are broad in nature and are the basic ethic. The seven fundamental canons are more specific than the principles and cover the wide range of responsibilities to the public, client, profession, and their own development as a professional engineer. The Code also contains guidelines which spell out in detail the various areas of moral responsibility of the ASCE member. Single copies of this Code of Ethics are available from ASCE headquarters.

C. THE NATIONAL SOCIETY OF PROFESSIONAL ENGINEERS

The NSPE is a national organization of professional engineers. Membership is open to registered professional engineers of any branch of engineering. In recent years the organization has opened its membership to nonregistered engineers but in a special grade of membership. NSPE has state chapters that meet frequently.

One of the major activities of NSPE is to promote licensing of engineers by written competency examinations. NSPE has been instrumental in educating state legislatures in the need for registration laws and has also been alert to the proper organization and operation of registration boards. Recommendations to the respective state governors of qualified engineers to fill the vacancies on Engineering Registration Boards are made by NSPE. These nominations are usually in cooperation with the other major engineering societies.

The NSPE is alert to legislation that affects engineers and has supplied qualified engineers to give testimony before legislative bodies on pending legislation related to technical matters. In the past years NSPE has been the most active of professional societies with regard to public affairs. However, other major engineering societies have increased their action and concern in this area of activity in the last decade. As society in general has become more "activist," so have the engineering societies. They have become much more concerned with societal problems. The work of the engineer affects the environment to a great measure and decisions with regard to development and environment must have the input of the engineer before correct decisions can be made.

The NSPE has had a Code of Ethics for many years. It has been active in making its membership aware of its Code. Its Code is very much engineering-practice-oriented. The Code as revised in January 1974 has 15 sections. The preamble has three statements; the first stresses honesty and devotion to employer, clients, and public, the second focuses on increasing the competence and prestige of the engineering profession, and the third requires the engineer to use his knowledge and skill for the betterment of society.

Section 11 of the Code of Ethics contains a ban against engineers entering into competitive bidding for engineering services. It states, "He shall not solicit or submit engineering proposals on the basis of competitive bidding."

This section of the NSPE Code of Ethics as printed below was declared in violation of Section One of the Sherman Antitrust Act by the U.S. Department of Justice in 1971. Unlike ASCE and the AIA[2] (see Appendix for revised Codes of Ethics), the NSPE decided to legally contest the judgment of the DOJ. The NSPE lost its case in the lower courts, and carried the litigation to the highest court in the land. The principal defense before the U.S. Supreme Court was that the code provision was a reasonable restraint of trade because competition among professional engineers was not in the public interest. The Supreme Court, in early 1978, rejected this argument and ruled that the Society's rules were anticompetitive. More details of this DOJ action are contained in Section E of this chapter.

SECTION 11

The Engineer will not compete unfairly with another engineer by attempting to obtain employment or advancement or professional engagements by competitive bidding, by taking advantage of a salaried position, by criticizing other engineers, or by other improper or questionable methods.

a. The Engineer will not attempt to supplant another engineer in a particular employment after becoming aware that definite steps have been taken toward the other's employment.
b. He will not pay, or offer to pay, either directly or indirectly, any commission, political contribution, or a gift, or other consideration in order to secure work, exclusive of securing salaried positions through employment agencies.
c. He shall not solicit or submit engineering proposals on the basis of competitive bidding. Competitive bidding for professional engineering services is defined as the formal or informal submission, or receipt, of verbal or written estimates of cost or proposals in terms of dollars, man days of work required, percentage of construction cost, or any other measure of compensation whereby the prospective client may compare engineering services on a price basis prior to the time that one engineer, or one engineering organization, has been selected for negotiations. The disclosure of recommended fee schedules prepared by various engineering societies is not considered to constitute competitive bidding. An Engineer requested to submit a fee proposal or bid prior to the selection of an engineer or firm subject to the negotiation of a satisfactory contract, shall attempt to have the procedure changed to conform to ethical practices, but if not successful he shall withdraw from consideration for the proposed work. These principles shall be applied by the Engineer in obtaining the services of other professionals.

As observed, the NSPE Code as contained in the Appendix is very detailed and complete in defining ethical standards in the practice of engineering. If all engineers would operate within the provisions of this Code, the environment of engineering employment would be very satisfactory.

D. ENFORCEMENT OF CODES OF ETHICS

All members, when they join ASCE, NSPE, AIA or other technical societies, agree to abide by the respective Code of Ethics of that society. This responsibility is usually pointed out in the membership application. ASCE has been vigorous in the enforcement of their Code of Ethics for many years. Infractions of the Code by members can be brought to the attention of the Executive Committee of ASCE. The Executive Committee then refers the case to the Committee on Professional Conduct for investigation. This Professional Conduct Committee investigates the charges thoroughly and quietly with no outside publicity. Upon completion of their investigation the Committee then reports their findings back to the Board of Direction. If they conclude that the person is guilty of violation of the code they will agree upon a penalty that should be imposed, considering the circumstances and severity of the infraction. The defendant can agree or not agree to this course of action. If he agrees then the committee reports the details of the investigation back to a closed meeting of the Board of Direction of the Society. This special meeting of the Board is called a Professional Conduct Hearing. The Board can agree with the Committee on Professional Conduct or decide if the penalty is too severe or too lenient. If the Board feels the penalty is too lenient they will require a full hearing on the case by the Board. At this full hearing the defendant is requested to be present at the hearing and he can bring legal counsel and/or witnesses in his behalf. The Board will then sit as a judge and jury with the Committee on Professional Conduct presenting their findings and with the defendant making any facts known which he desires for pleading his case.

The Board, in closed door session, deliberates the case and decides if the defendant is guilty or not guilty of infractions against the Code of Ethics. If guilty, the Board will decide the penalty. Since the only power the Board has is memberhip status, the Board will decide upon one of the following penalties:

1. a letter of reprimand to the guilty member or members with no official disclosure to the membership of the names of the defendants;
2. a letter of reprimand with disclosure to the membership of the action of the Board and names of the guilty parties;
3. dismissal from the society for a given number of years;
4. expulsion from the Society with no privilege of readmission at a later date.

In the case of actions 2, 3, or 4, the membership is notified of such action through publication in *Civil Engineering,* the official magazine of ASCE. The Society can also release their findings to the proper authorities when the conduct of the guilty parties has been in violation of the civil or criminal laws of the state. In cases of fraud, bribery, or embezzlement this action would be taken.

In the past history of the ASCE, the Society has been vigorous in enforcing its Code of Ethics. This author sat in on several cases of professional conduct when he served as a member of the Board of Direction of ASCE. He can verify that all cases were handled fairly, seriously, and vigorously. The welfare of the defendant was always considered as well as the health of the Society. The Committee on Professional Conduct performed their assignments thoroughly, and many man-hours were expended by the Committee and the Board on each case. Board members did not look forward to conduct hearings but nevertheless took their responsibilities seriously.

At the surface level the student or young engineer may think that just a reprimand or loss of Society membership is not much of a penalty for the Society to impose. In almost all cases the defendants value their membership and reputation to such a degree that they will go to considerable effort and expense to appear before the Board and plead their case and disclose any extenuating circumstances, asking for as light a penalty as the Board may consider.

A majority of cases tried by the ASCE have been infractions involving relations between engineers and between engineers and clients. However, in recent years cases involving public trust, such as bribery and influence peddling have become infamous. Some of these cases will be discussed in Chapter 4.

As previously described, the only means of enforcement of codes of ethics of professional groups is by expulsion from membership and the resultant bad publicity. In some instances it may be possible to bring action to withdraw the professional license. This latter action has to be processed through the courts with the state registration board as the body to bring the complaint to the responsible authority. The majority of the members of the state registration boards will usually be members of the profession of the accused so that action can easily be brought. However, as pointed out, there must be a clear case of willful misconduct before the courts will recall a professional license. The state registration acts must also be written so that wrongful acts such as bribery and payoffs as well as incompetence are grounds for withdrawal of professional license.

Most offenders of professional societies' Codes of Ethics, when faced with an accusation, have indicated a great desire to retain their membership. The resultant bad publicity can have a very detrimental effect on the ability of the offender to obtain new professional engineering contracts. If the offender is an engineering employee and not in private business, the record of misconduct will travel with him for many years. Most reference forms for new employment will ask, "do you know of any action of the applicant that shows a deficiency in character." Unless there is knowledge that a character weakness has been overcome, there is a reluctance to hire someone with a past history of conduct not in keeping with a Code of Ethics.

It should be understood, however, that the main purpose of a Code of Ethics of a professional society is not to serve as a means of removal from membership those who do not conform. The main purpose is to serve as a guide and an emblem of moral standards for which the profession subscribes. It is an "ethic" for the education of students and members just entering the profession. To serve this latter purpose there must be programs within the engineering colleges to present and discuss Codes of Ethics and the need for high integrity of the members of the profession.

E. ACTIONS OF THE U.S. DEPARTMENT OF JUSTICE AGAINST CODES OF ETHICS

As indicated earlier the American Society of Civil Engineers has had a Code of Ethics since 1914. There had been no question of the legality of this Code for over fifty years. Members had been held accountable for conformance to the articles of this Code and relatively few had been expelled for infractions. However, in the summer of 1971, the U.S. Department of Justice informed ASCE, NSPE, and the American Institute of Architects (AIA) that it was serving these societies with a Civil Investigative Demand requiring documentation relating to the adoption, administration, and enforcement of the article in their codes of ethics prohibiting their members from submitting priced proposals under conditions that constitute price competition for professional services. All three societies had similar provisions in their Code of Ethics.

This prohibition was Article 3 of the ASCE Code of Ethics, and Section 11 of the NSPE Code.

In order to fully comprehend Article 3 in the ASCE Code of Ethics and also similar articles in the code of ethics of NSPE and AIA, one must be aware of the full concept and operation of the design contract procedure. It will be briefly discussed here and in more detail in Chapter 8.

The ramifications of a design contract between the future owner of a facility and the firm which will design it are quite different from those of a contract between the owner and builder. In the latter case, there are definite plans and specifications which detail the work to be done down to the last nut and bolt. In the case of the contract between owner and designer, the work to be done by the designer can only be defined in broad, general terms. For instance, a facility is to be designed such as a water treatment plant, bridge, or industrial plant. The final functional capacity and location of the project can be defined. However, the degree of expertise and optimization in the design, and the overall knowledge and experience of the designers cannot be specifically defined and set forth in a contract document. The engineering profession, therefore, believes that the design engineers should be selected first on the basis of their experience and proven ability to perform the design task. The fee for doing the design is then negotiated with the selected firm.

The result of selecting the design engineer on an open bid basis is that the firm which presents the lowest bid will most likely receive the contract. Such procedures can and most likely will result in quality of work below optimum. Approximations can be taken in the design calculations and man-hours of engineering saved by duplication with past projects. The nature of the design with least-cost engineering will be that which has been done before, since the second or third time through, the design can be completed in fewer man-hours and thus the designer is more likely to meet his low bid with his "standardized design." When designers are forced to quote low prices for design work because of the open bid process, the resultant design is most likely to be less than optimum. Therefore, the owner will be paying more for the facility than if he had a more precisely designed product. This extra cost in facility may be considerably more than the saving in design cost by the competitive bid procedure.

The above explanation of results of selecting engineering work on the basis of price in a bidding context was the reason ASCE, NSPE, and AIA prohibited their members from taking part in the competitive bidding process. The main rationale was that in the environment of competitive bidding, inferior engineering would take place which in the end would result in a more costly facility and in addition, innovative, progressive engineering design would be suppressed.

Irrespective of the quality of engineering that may result, the Justice Department in January of 1972 informed ASCE and the other two societies that it was planning to initiate a suit against the Society under Provision 1 of the Sherman Antitrust Act. The suit would allege that Article 3 (the prohibition against price bidding for engineering contracts) in the Code of Ethics of ASCE was a restraint of trade. Similar provisions against bidding in the codes of the other two societies were also targets of the DOJ.

In mid 1971 after the first approach by the DOJ, ASCE had sought legal counsel and had been advised that under the circumstances and under the law it would be best to delete Article 3 from the Code of Ethics. The Boards of Direction of ASCE as well as NSPE and AIA were faced with three alternatives:

1. To voluntarily remove Article 3 from the Code.
2. To await the filing of a suit and then to seek to negotiate a consent decree.
3. To await filing of a suit and then contest it in the courts.

After intense deliberation the Board selected the first alternative. If this action would satisfy the DOJ, the Society would no longer be able to expel those members who entered into the action of submitting priced bids as it had done in the past, but it would not prohibit ASCE, in an educational program, of informing potential clients of the negative results of priced bids for engineering services.

Even though Article 3 was removed from the Code of Ethics effective on January 1, 1972, the DOJ did file a civil antitrust suit against ASCE. ASCE and the DOJ did enter into a consent decree in which terms of the decree were negotiated. Under this procedure, ASCE accepted an injunction, but without admission of guilt for any antitrust violation. The terms of the consent decree were that ASCE would amend its Guides to Professional Practice, Codes of Ethics, manuals, rules, bylaws, resolutions, and any other policy statements; to eliminate therefrom any provision which prohibits or limits the submission of price quotations for engineering services by members as unethical, unprofessional, or contrary to any ASCE policy. It also prohibits ASCE from adopting or disseminating any such provision, statement, or implication in the future.

ASCE agreed to the final judgment without trial or adjudication of any issue of fact or law and without final judgment constituting evidence or admission by any party with respect to any issue of fact or law.

The final judgment did not prohibit ASCE from advocating or expressing its view that the procurement of engineering services involves consideration of factors in addition to a fee. The DOJ also confirmed the right of ASCE to engage in legislative activities seeking federal, state, or local action which would prohibit the selection of engineering services on the basis of competitive bidding. Several state legislatures and the federal Brooks Bill relating to the procurement of engineering services by governmental agencies set forth procedures for the securing of engineering services other than through competitive bidding.

There was considerable difference of opinion on the part of the membership of ASCE as to whether or not ASCE should accept the consent decree. The alternative would be to go to the courts to prove that Article 3 and the manner in which ASCE had enforced this article was not contrary to Section 1 of the Sherman Antitrust Act. This route would require considerable expenditure of funds for legal fees. Such funds would have had to come from increased dues. Another very important factor in the decision not to fight the DOJ suit was that if a final ruling rendered by the courts was adverse to ASCE, this ruling could be used by public agencies and private persons to bring treble damage actions against the Society, including its officers, directors, and any members who had in the past refused to furnish priced proposals. It was the feeling by the Board of Direction of ASCE in 1972 and upon the advice of legal counsel that the risk was too great; therefore, the consent decree was signed and since 1972, Article 3 has not been part of the ASCE Code of Ethics.

NSPE opted to contest the action by the DOJ. This suit has now (1979) gone the entire route through the courts to the U.S. Supreme Court and NSPE has lost in each court decision. The exact final court settlement has not been reached. The decision of ASCE to enter into and accept the consent decree must be considered justifiable action.

An interesting fact to the subject of this case is that doctors and lawyers did not run afoul of the law (antitrust act) since neither of these professions had ethical prohibition against competitive bidding. It is quite unthinkable to the public that one would ask for competitive bids from doctors or lawyers when the need for medical or legal help would be necessary.

The AIA followed in the same path as ASCE and deleted any prohibi-

tion of competitive bidding. The AIA has also been active in the education of would be clients in the folly of selecting an architect on the basis
of lowest price only.

With the acceptance of the consent decree, ASCE officers considered
that the modified Code of Ethics as of January, 1972 would prevent any
further conflict with the United States Department of Justice. This was
not to be so. In 1973 a joint venture of two engineering firms was
entering into negotiation with the Metropolitan Water Works of Bang-
kok, Thailand to provide engineering services for a five-year water
supply construction contract. It was claimed by the joint venture that
just before the contract was to be signed, another joint venture of an
American firm and also a Thai firm submitted a proposal lower than the
firm under consideration. Because of this last-minute lower price, the
second joint venture was given the contract. This brought a charge that
the principals of the American firm entering the lower price had violated
the ASCE article in the Code of Ethics that "deems it unprofessional to
attempt to supplant another engineer in an engagement for engineering
services after definite steps had been undertaken towards his employment." The Board of Direction of ASCE heard the charge against two of
the major officers of the firm (both ASCE members) and found them
guilty as charged and suspended them from membership in ASCE for
three and two years, respectively.

Upon this action the DOJ again appeared on the scene and charged
ASCE in contempt of consent decree of 1972. ASCE officers did not
believe that their suspension of the two members was in violation of the
consent decree, since it was not due to submission of a priced bid but
rather to their attempt to (and successfully) supplant another engineering firm after they had, for all practical purposes, been selected for the
project.

The first move of ASCE was to consider legal opposition to the contempt charge, but upon further consideration and legal advice that a
court case would be hard to win, ASCE agreed to drop the suspension
charges and reinstate the two members. It was then considered quite
necessary to look at a complete revision of the ASCE Code of Ethics.
This change in the Code of Ethics was made and adopted by the members
on September 25, 1976, and became effective January 1, 1977. This new
Code is reproduced in the Appendix of this book.

Until the 1970s, professional societies were not bothered about their
codes of ethics from any legal direction. It was considered almost unthinkable that a learned society could be in violation of antitrust laws.

It was believed that professions did not engage in interstate commerce—a requirement that must be proven in federal antitrust suits. In 1975 the famous Goldfarb decision was reached by the U.S. Supreme Court. This case involved the legality of a minimum fee schedule recommended by the Fairfax County Virginia Bar Association. It was ruled that the fee schedule was in violation of the Sherman Antitrust Act. This landmark decision gave notice to the professions that they had no immunity to antitrust laws. The Court was able to justify the ruling that lawyer fee schedules could affect interstate commerce. Interpretation of civil laws by courts and in many cases by lawyers in departments of justice have established what is right and wrong. Ironically, federal and state departments of justice have made decisions to weaken ethical constraints of professional societies in the post-Watergate era when such societies have seen a compelling need to stengthen the ethical climate of their professions.

It is now apparent that even Codes of Ethics can run contrary to the law—i.e., the law as interpreted by legal persons. Engineering societies now realize the conflict that may arise with any rules or regulations that may in any manner attempt to control the "business operations" of any of its members or the society in general. How, then, can a learned society control the ethical actions of its members for the good of the country and its citizens? This does present a challenge to the engineering professions. Codes of Ethics have diminished influence without a means of enforcement, especially if any act of enforcement is judged by the courts to be in violation of the laws of free trade.

One result of these conflicts between learned societies and departments of justice is that policies and procedures relating to codes of ethics or disciplinary action against any member have to first be carefully scrutinized by legal counsel.

F. GUIDELINES TO PROFESSIONAL EMPLOYMENT

Professional employment should be more than just a job. As a teacher for many years of prospective engineers, the author has tried to instill in his students a feeling of professionalism and a belief that engineering is full of exciting and demanding challenges. It is hoped that this will be the case for all those young people who take their diploma in one hand and a calculator in the other and go forth with great expectations.

In 40 years of engineering practice, among the dozen employers the author has worked or consulted with, there has been a wide range of character. Some firms had high morale while others were disappointing. Poor employee–employer relations are expensive to an engineering firm. An organization that sells engineering services or products that depend on creative engineering lives or dies with the quality and productivity of its engineering staff. No company can afford to have unhappy, discouraged engineers on its payroll.

Mutually satisfying relationships between employee and employer can come about only by both parties fulfilling their ethical obligations to each other. Each has a responsibility and need for knowledge and understanding. Good employee management does not just come naturally, but basic personality has a major influence. Management training is very much a part of the business world today. The larger firms are apt to be more involved in management training than smaller organizations, although this is not always the case.

Professional societies have recognized the need of good employer-employee relationships in order to maintain a healthy, strong, and progressive profession. The American Society of Civil Engineers has published a manual entitled "ASCE Guide to Employment Conditions for Civil Engineers." This is very similar to the booklet "Guidelines to Professional Employment for Engineers and Scientists" [3]. This booklet and the ASCE manual is endorsed by over 28 separate professional engineering societies. Copies can be secured from any one of the endorsing societies. The "Guidelines" booklet was copyrighted by Engineers Joint Council in 1978. The first edition was dated January 1, 1973. This indicates that the real concern for employer-employee relationships is of recent vintage.

The "Guidelines" has four basic topics. They are: (I) recruitment; (II) employment; (III) professional development; (IV) termination and transfer.

The chapter on recruitment has some guides to college students who are paid expenses to visit a firm for a job interview. If multiple interviews are held during one trip then all the companies should share in the cost of the trip. Gaining employment on the basis of divulging trade secrets or proprietary information is unethical on the part of the employee and the new employer. However, when an engineer changes employment he certainly brings with him the knowledge and experience gained in all previous positions of employment. The principle set forth by the "Guidelines" is that an engineer should not be hired just to gain information from a competitor.

A major responsibility of an employer in recruitment is to provide the prospective employee with all the information that will have a bearing on the desirability of accepting employment.

The "Guidelines" stress that employment of engineers should be consistent with ethical practices of the profession. The loyalty of employee to employer is part of the Codes of Ethics of all engineering societies, but this loyalty should be within the bounds of ethical practices. Cases where loyalty extended into unethical and illegal practices will be given in Chapter 4.

Employment in technical fields may lead to development of new devices that can be patented. There should be a clear company policy (in writing) with regard to patent rights. A prospective employee should be given a copy of this for his reading before he accepts employment and he should keep this dated and signed company policy as a record of his contract with his employer. This can forestall legal problems when a question of patent rights develops. If a patent develops as a result of work within the company and during the course of working hours, then there should be mutual rights to the patent. This, of course, assumes that the employee contributed to the development with his mind and skill. Patents developed outside the working hours of employment and not related to the work of the employer should, under most cases, be the property of the employee. However, it may be difficult to determine if the work of the employee contributed to the development of the patent even though away from work. There have undoubtedly been many legal cases covering many circumstances in which decisions have been handed down by the courts. There isn't enough space to look into this phase of law. The idea to be set forth here is that all questions of patent rights should be handled in an honest, ethical manner and that agreements should be made before any patents develop.

A subject that is addressed in Section II of the "Guidelines" is that of titles. Titles denoting professional engineering status should be restricted to graduate engineers or those holding a professional engineering license. The first company that employed the author in his engineering career was full of employees who called themselves engineers, but had had little formal engineering training. They had moved through the ranks as draftsmen, and in most cases their judgment and ability as engineers created many difficulties because of lack of a sound technical foundation. This, together with time clocks and a generally poor professional environment, encouraged employment elsewhere. It wasn't long thereafter that the engineering activity of the company closed down.

A possible source of friction between employer and employee is overtime work. There should be a clear company policy with respect to overtime, and when the employee is not paid overtime, there should be time off during slack periods. If the overtime is for extended periods of time then the employee should receive compensation for his time. The employee should not feel that his job is strictly an eight hour day, and if it becomes necessary to occasionally work an extra hour or two to meet deadlines, he should be willing to cheerfully do so without thought of extra pay. To a professional, the successful completion of the job in the required time period is of great importance.

The employer should see that the office environment is conducive to mental well-being. Desks, tables, chairs, etc., should be adequate and of good quality, and should present a professional appearance. Poor office planning can result in inefficiency and a decrease in employee morale.

A fault of many consulting engineering firms is the failure of management to keep the employees, even the professional staff, aware of the work status. Short periodic meetings to inform everyone of work progress, pending projects, and the general situation of the company will give the employees a feeling of belonging, a sense that management considers them an integral part of the organization. It helps greatly to dispel office rumors, which are a waste of time and destructive to positive mental attitudes.

Many companies have determined that profit sharing plans are very beneficial to employee efficiency, employee retention, and in developing a spirit of loyalty among the employees.

The major asset of companies selling engineering services or engineered products is the engineering staff. A truism of engineering management is that when a firm's operations and policies deteriorate, resulting in low morale, the best qualified people are first to leave. They are the employees whose skills are most marketable and who have pride in their work and concern for their future. A very human trait is the desire of recognition for good work. The employer should reward those whose performance adds to the financial well-being and professional reputation of the firm. Ethical considerations require that careful attention be given to the contribution of all members of the firm when monetary rewards are distributed.

The third subject of professional employment in the "Guidelines" is the professional development of the engineer. It is the responsibility of the engineer to maintain technical competence. A college degree is just the commencement of a professional career. Much learning is required beyond that point in life. This will require membership in technical societies, reading of current literature, and attendance at professional meetings. Enrollment in continuing education courses at different times may be necessary. All of the mentioned activities will help toward the goal of professional registration.

The engineer should be looking toward using his talents and sharing his knowledge through public and professional service. He can do this by committee assignments, holding office in technical societies, and the presentation of papers and talks. The entire profession of engineering will benefit from the high standards of service its members render to society.

The employer should cooperate with the employee in the development of the engineer as a professional and as a contributor to society. The employer should encourage membership and activity in professional societies. Time and money should be budgeted by the company for assistance in the development of young engineers and support of older professionals in continuing education and professional social activity. Encouragement by management in the preparation and presentation of technical and professional papers should be standard company policy.

The fourth section of the "Guidelines" covers termination and transfer. There are definite ethical procedures on the part of both employee and employer at the time of termination of an employee. Adequate notice should be given by either party that initiates the termination. The "Guidelines" recommend that permanent employees should be given notice of one month plus one week per year of employment prior to termination or equivalent severance pay. If an employee is transferred then all costs of transfer should be borne by the employer.

There are a number of additional topics covered in the "Guidelines." Adherence by both the employee and employer to the rules and procedures will develop a high ethical environment in the office and *will* enhance the professional reputation of both employee and employer.

NOTES

1. Changed to Accreditation Board for Engineering and Technology (ABET) in January, 1980.

2. The American Institute of Architects' Code of Ethics is under review and revision at the time of printing of this book. Copies of the latest code may be obtained by students, teachers, or professionals by writing to American Institute of Architects, 1735 New York Avenue, N.W., Washington, D.C. 20006.

REFERENCES

1. W. H. Wisely, "Public Obligation and the Ethics System," American Society of Civil Engineers, Preprint 3415, October 1978.
2. C. R. Schrader, "Professionalization—and a Referent Code of Ethics," *Engineering Issues*, ASCE, October 1974.
3. "Guidelines to Professional Employment for Engineers and Scientists," Engineers Council for Professional Development. (Copies available from major engineering societies.)

4. MISCONDUCT OF ENGINEERS AND ASSOCIATES

A. INTRODUCTION

In the previous chapters the concept of moral standards and ethical behavior in professional societies has been developed. A presentation and discussion of the codes of ethics has been made. It is hoped that the presentation made in these chapters will assist the engineer, and especially the student engineer, in helping him decide what is ethical and right in professional practice. It is hoped that each engineer, as he enters the profession, will have developed his own set of ethical standards that will not only keep him from being in conflict with the Codes of Ethics of his professional society but will also build his reputation as an honest and moral person.

This chapter will present some case studies of unethical behavior of engineers that has occurred in the past. These cases have been tried before the engineering societies as professional conduct cases and penalties imposed. Many of the cases have also been tried or investigated by the constituted legal authorities. In some cases prison terms have resulted from unethical activity.

The latter part of this chapter will contain some hypothetical ethical situations that could and most likely have developed. It is hoped that classroom discussion in light of ethical standards previously presented will take place. In some situations it may be difficult to draw a guilty or not guilty verdict. Every case has its special situations with a wide variety in the background of participants. Many believe that a person is not guilty if he did not know better, while the courts say that ignorance of the law is no excuse.

It is hoped that ethical training will become part of the education of every professional, who thereby learns what is expected in the world of professional practice. Ideally, moral training will begin at an early age in the home and in religious activity—it is best learned while a person is young. Nevertheless, education in the ethical behavior of professional practice can and should be part of the formal training of the student engineer.

B. CASE STUDIES OF UNETHICAL BEHAVIOR

A major source of illegal and unethical behavior is the selection of engineers and architects for public projects. The scenario for such a selection involves the following: (a) a project involving large sums of money for construction and considerable amounts for engineering or architect fees; (b) a public official or officials who are modestly paid (or maybe even poorly paid); (c) an engineering or architectural firm that needs work in order to remain in business; (d) a contract selection process whereby the public officials can make the selection in somewhat secret circumstances.

The process is that the public officials or official makes a selection of the firm who will be awarded the design contract. This selection should be on the basis of the best qualified firm which can expeditiously prosecute the work for a reasonable fee. Within any geographic area there are usually a number of firms which appear to the officials to be adequately qualified. The door is then opened to the firm, which to a fair degree at least meets the above requirements, but which has been generous in political contributions to the public officials involved. In most cases there is no actual breaking of the law if the legal limitation on campaign contributions has been observed. This restriction has been circumvented by family or friends also making generous contributions.

From the position of awarding contracts to generous contributors, it is only one step further to make awards with the agreement that a certain percentage of the engineering or architectural fee will be paid back, in cash, to the public official. This action is, of course, illegal everywhere within the United States and in most other countries. Nevertheless, this type of illegal and unethical action has taken place in the past and is possibly currently happening. This type of case with other cases is described in detail below.

Case 1: This case was a long-time operation in the state of Maryland. Kickbacks were made to several public officials including a county and state official by Maryland engineering and architectural firms from 1962 to 1974 for receiving design contracts. This whole sordid case has been reported in detail elsewhere [1] and in the investigations that took place, several engineering and architectural firms in Maryland admitted to giving bribes and kickbacks to a number of public officials holding county and state positions. In the ensuing trials several persons were convicted and fined as well as given prison sentences for their action. Spiro T. Agnew, who was the governor of Maryland during part of this time period and was accused of having received some of the money, resigned from the Vice Presidency of the United States on October 10, 1973. Several involved members of ASCE and the Maryland Society of Civil Engineers were expelled from their respective societies. This whole complex of bribery and kickback cases was the most notorious in several decades. It resulted in a "black eye" for the engineering profession as well as the political environment of the entire state of Maryland.

Case 2: A New Jersey consulting engineering firm had a contract to provide engineering services that included field and office engineering. Since the project was complex and the amount of engineering man-power was indeterminate at the signing of the contract, the award was made on the basis of a fixed fee per man-hour of engineering. This required each engineer in the firm to record on a time sheet the hours he worked on the project. The time record was to be audited periodically by the governmental agency.

One of the partners in the firm instructed a departmental head who in turn instructed a subordinate to have some engineers who were working on a separate fixed-fee contract charge their time to the government contract. The department head and his assistant wondered about the illegality of this action and even questioned the partner. His reply was that it was just a bookkeeping procedure and in the end it would all even out. These two engineers were born and educated in a foreign country and were unsure of the business procedures in the United States. They trusted the head of the firm and had the individual engineers charge their time to the wrong project number. Government officials eventually suspected the wrongdoing and in the following legal investigation both of these directed employees confessed to the wrong action and became witnesses against the firm and especially the partner. The resulting trial led to a conviction of fraud for the partner and a jail sentence. The two subordinate engineers were not indicted but legal fees in presenting their cases in the grand jury investigation were costly. The firm was dissolved as a result of this illegal action. All three engineers' ethical conduct was set for trial by ASCE Board of Direction. The partner resigned from ASCE before the hearing and his resignation was received with prejudice (no future membership without a hearing before the Board). The department head had a professional conduct hearing and received a one-year suspension from ASCE. The other engineer received a letter of reprimand from the Board of ASCE. Both members were found guilty by the Board of ASCE of violation of the ASCE Code of Ethics in that they acted in a manner derogatory to the honor, integrity, and dignity of the engineering profession.

Case 3: This case had a history of traveling down the primrose path. An engineering consulting firm solicited the county commissioners of a given county for the design contract of a water supply system. They were told they had the job but a final signed contract could only be forthcoming after a future bond election, which if passed would provide the funds for the project. They were encouraged to go ahead with the engineering design work since it was obvious that the project was absolutely necessary and that there was no real opposition in the county to the project. Work in the engineering office was slack and the county officials wanted an early start so the engineering firm began work on a verbal understanding only, and had expended about $50,000 for the engineering when the bond election was held and the voting was affirmative. Upon approaching the county officials, they were told they would receive a written contract upon a verbal agreement to kick back a percentage of the engineering fee to the county commissioners. The engineering partners were informed that this was standard procedure in the county. Refusal would mean they would not get paid the $50,000. This would probably require bankruptcy, since they had been operating on borrowed money. The county attorney was probably involved in the whole scheme, so that if the engineers attempted an exposure they would get the brushoff as well as possible reprisals. In the darkest hour of their professional careers they decided to go along with the payoffs. At a following election the county commissioners involved were not re-elected. The engineers were relieved since the work was still in progress. However, when the new officials took office they demanded and received the same kickbacks. This illegal procedure appeared to be a way of life in this county. The downfall came when the federal district attorney obtained information that all was not legal with regard to the project. Part of the cost of the project was borne by the federal government. As soon as the F.B.I. began their investigation, the engineering firm turned state's evidence and cooperated fully with the district attorney. The federal judge who tried the case said that without the voluntary cooperation of the engineering firm a case against the county commissioners could not have been successful. At the trial, evidence was presented that several partners of other engineering and architectural firms had been involved in similar kickbacks. There was no indictment against the engineering firm because of their cooperation with the prosecution. Several county officials were sentenced to federal prison.

The ASCE Board of Direction conducted a hearing on the ethical practices of the partners of the firm and found them guilty of violation of the ASCE Code of Ethics. They were sent a letter of reprimand. A more severe sentence was not voted since the action of the partners of voluntarily turning all information over to the district attorney was greatly in their favor.

Case 4: The chief executive officer of an engineering company made contributions totaling $10,000 to a national presidential campaign and had pleaded guilty to violating the federal law prohibiting corporate contributions to federal election campaigns (Section 610, Title 18 of the U.S. Code.) It may appear that he was making a personal contribution and not a corporate gift. However, the company had a plan whereby funds would be accumulated for contributions to civic organizations and political campaigns. Company individuals were given bonuses and after deduction for personal income taxes, the balance would be placed in the fund. These funds were then used for civil and political contributions by the company. The legal verdict in court was that this procedure was for the purpose of circumventing the spirit of the law.

The ASCE Board of Direction heard the evidence against the chief officer and found him in violation of the Code of Ethics and he was sent a letter of admonition. The action of the Board was reported in the official publication of ASCE.

Case 5: This case involved a different type of ethical problem. It involved a charge of plagiarism and breach of copyright with respect to the publishing of a manuscript in an ASCE journal.

The person was charged with violation of the Code of Ethics in that "it shall be considered unprofessional and inconsistent with honorable and dignified conduct and contrary to the public interest for any member of the American Society of Civil Engineers to act in any manner derogatory of the honor, integrity, or dignity of the engineering profession.''

Evidence was presented to prove that the person charged had published a paper which was almost an exact replica of lecture notes in a class in which he had been enrolled. These notes were from sections of the class professor's Ph.D. doctoral dissertation. The thesis had been published several years earlier as part of the proceedings of an international congress. This material was not known by the engineers who reviewed the questioned paper for Society publication. No credit was given by the author of the paper to any of the professor's earlier work.

After presentation of all evidence, the Board of Direction of ASCE voted to suspend the accused from membership in the Society for a period of two years.

Case 6: This case was nearly the exact opposite of the previous case. Here the professor used the work of two former students in the publication of two articles in two different Society journals under his own name without acknowledging the work of the students. The resulting investigation showed that the students' material had been used almost word for word in the professor's papers. The students also indicated that they had received very little help from the professor in preparing their papers.

The professor, in defense, indicated that he had been invited to give a talk at a society meeting on the same subject as the students' term papers. He gave the talk using notes and visual aids. Later he was asked for a written version of his talk and he submitted the students' papers under his own name. He claimed he did not know that the material requested was for publication. However, he again submitted the students' papers, slightly modified, to a second society for publication.

The ASCE Committee on Professional Practice, which investigates all cases of alleged misconduct of ASCE members, recommended to the Board of Direction that the professor be suspended from ASCE for a period of five years. However, before the Board hearing took place, the professor had been dropped from the Society for nonpayment of dues. He was informed that the evidence of unethical behavior might be considered with any future request for membership reinstatement.

Case 7: This case shows a poor employee–employer relationship. It may be typical of similar situations that have taken place in several engineering organizations.

In this case an engineering employee of a consulting firm was project leader on several projects. He decided to leave the consulting firm and persuaded several of the clients with whom he had been closely working to cancel their contracts with the consulting firm and then enter into agreements with him to complete the work. The question of whether the solicitation took place before or after the employee left the consulting firm was debated by the consulting firm and the employee. The employee contended that he left the firm (without notice) because the firm had not made good on certain promises relating to employment conditions. He further contended that the clients requested that he take over the projects after he left the firm since he was the person they had worked with and who knew the most about the projects.

The complaint by the consulting engineering firm was first made to the state engineering society. The state society made an investigation and found the engineer guilty of a breach of ethics and issued an admonishment. About five years after the first allegations were made, a complaint was made to ASCE. The ASCE Committee on Professional Conduct made an investigation, and after compiling a considerable amount of information, it could not clearly determine if the accused was guilty as charged. The employee did supplant his former employer as the engineer for some of the previous clients. However, it could not be clearly established in the time frame given as to when he solicited the assignment and if he initiated the switch or the clients did. CPC found it difficult to evaluate the various charges and countercharges. However, CPC was critical of the conduct of both parties during the controversy and voted to issue a mild letter of censure to each.

This case brings forth two subjects in the discussion of ethics. The first relates to an engineer forming his own firm after being in a responsible position with another firm for several years. Ambitious and capable engineers are likely to have such a goal in their career plans. In the course of his employment, an engineer will probably have formed good relationships with several of his firm's clients. In organizing a new firm, the prospect of immediate contracts can be a major incentive to make the break from being an employee to that of an employer. Should he solicit the clients of his previous employer for future engineering work if he starts his own firm? It would not be ethical to solicit business in competition with one's own employer. Once an engineer establishes his own firm then he can solicit engineering contracts from any possible source. However, it would be in violation of ASCE Canon 5a to attempt to supplant a firm which has already been selected by a client.

Case 8: A building structure was under construction by a contractor for a county agency. It had been designed by a consulting firm. Cracks developed in the concrete structure and the county agency and design firm had questioned the contractor's work. Complaints had been filed by both the county agency and the contractor. The contractor then hired an independent consulting firm to prepare a report on the cause of cracking and an analysis of the contractor's performance. In the resulting claims the design firm felt that the evaluating firm had attempted to falsely and maliciously injure the professional reputation of the design firm. There was the charge that the second consulting firm was reviewing the work of another engineer for the same client without the knowledge of the design engineer.

The Committee on Professional Conduct of ASCE investigated the charges and found that the second consulting firm, which was hired by the contractor, had not attempted to falsely or maliciously injure the professional reputation of the design engineer. In the review and evaluation of the project, the second engineer had been hired by the contractor and not by the county agency. He was to primarily review the work of the contractor and not that of the design engineer.

The second consulting firm was not found guilty by CPC and the charges were dismissed.

The problem of a second engineer hired by a client to review the work of the first engineer is covered by item 4.h of the ASCE Code of Ethics. It is not uncommon that a client may not be entirely satisfied with the design, evaluation, etc. of an engineering firm. It may wish to obtain a second opinion. The proper procedure would be to seek the approval of the original firm before the second firm was hired or told to proceed with the evaluation. If this approval was not forthcoming, the client could dismiss the first firm and hire a second firm. The second firm could accept the assignment without breaking the Code of Ethics.

It is common in European practice for the client to hire a second engineer, separate from the design engineer, to review the design of complex major structures. The second engineer is called a *proof engineer*. He is considered an authority in the specific design field. The design firm knows their work is going to be reviewed so there are no unknowns in this aspect of the project. There is no violation of a professional Code of Ethics in this procedure.

Case 9: A contract was to be let for the design of a multimillion-dollar public facility. Several firms were necessary to do the work because of the size and the contract was let to some associated consulting firms of engineers and architects. However, before the contract was to be let, the firms were solicited for substantial contributions to an election campaign. The firms agreed and made the contribution prior to signing a contract. Some time after the contributions were paid, legal authorities became aware of this action and a grand jury handed down indictments to several individuals who had made the contributions.

ASCE did not take action until the court cases were settled. This is the usual procedure since the trial will bring forth testimony that will be of importance in the investigation of violation of the Code of Ethics. A professional society does not have the power of subpoena and can only obtain evidence from witnesses if they are willing to testify.

During the court action one of the consulting firms pleaded nolo contendere (no contest of the indictment). The ASCE Committee on Professional Conduct obtained transcripts of the court proceedings and interviewed persons involved in the legal charges. All the accused ASCE members admitted that political contributions had been made but did not feel this was an ethical wrongdoing. They believed they had made a legitimate political contribution and that they would have been awarded the contract even if they had not made a contribution. However, they did admit that they had not received a signed contract until they had made the contribution.

Court testimony showed that there was a definite pattern of kickbacks to the elected official's henchmen and that the amount paid by individuals was directly related to the percentage of ownership they had in their respective consulting firms. All the money was delivered in cash and in plain envelopes. No receipts were ever given. This pattern of contribution had never been made by the individuals involved before this contract was consumated.

The Board of Direction of ASCE heard the case and found the accused members guilty of exerting undue influence in the solicitation of engineering engagement and that they acted in a manner derogatory to the honor, integrity, and dignity of the engineering profession. This conduct was in violation of the ASCE Code of Ethics. The guilty parties were suspended from ASCE for a period of years.

The information given in the above professional conduct cases was supplied by ASCE without identification of persons or locations, except that Case 1 was reported in Reference 1 and has been referred to in many other publications. This case received national attention at the time the public charges were brought against former Vice President Agnew. ASCE has had other professional conduct cases, but this sampling was selected since they show a variety of unethical conduct in violation of the ASCE Code of Ethics as well as the laws of the land.

NSPE and other technical societies have had professional conduct cases and have conducted hearings in a similar manner to the ASCE. The author chose ASCE case studies since they were readily available and also because he was acquainted with some of the specific cases.

In the remainder of this chapter several situations will be presented that could possibly confront practicing engineers and students. They are given for the purpose of class discussion or individual evaluation. The Codes of Ethics presented in Chapter 3 should be used as a guideline. Every ethical confrontation will have its own particular circumstances that will certainly have a bearing on the course of action a particular person may follow.

> *Problem* 1: A student is nearing graduation and is offered a trip by a large consulting firm to a distant city to appear for a personal interview for possible employment. The firm has offered to pay the student's expense for the interview. The student knows there are several other firms in the same city that may possibly offer employment in his area of interest. His thinking is that he will stop in and visit these other companies while he is visiting the city for the interview with the first company. Is there anything wrong with this proposed course of action?

> *Problem* 2: A student has been offered a trip for an employment interview by a company located within an hour's drive of his home although several hundred miles from his university. He decides that Easter vacation will be a convenient time and since he was planning a trip home for this vacation from school anyway, this will save him the personal expense of the trip home. Should he accept the travel money from the firm considering that he was planning the trip home? Would it make any difference in the ethics of his action if he were serious or not about employment with the company? Would it make any difference if he accepted employment with the company? What is the suggested course of action for the student?

> *Problem* 3: A student has received offers of interviews from two companies several hundred miles from his campus but only fifty miles apart. The nearest airport to both firms would be the same. He considers the fact that he can see one firm in the morning and one in the afternoon and spend only one night in the hotel at the airport.

The thought occurs to him that he can interview with both companies and then bill both of them for the plane fare and hotel room. He can make a Xerox copy of the fare receipt and hotel bill for one of the firms. If any questions are asked by the firm receiving the Xerox receipts he can say that he lost the original and sent the Xerox copy from his own files. What is the situation ethics in this case?

Problem 4: A student is in his last semester of college. His department chairman calls him in and tells him that a local firm wants to hire a prospective graduate for part-time work with a view to an offer of permanent employment upon graduation if the student's performance is satisfactory. The student has the time to work since he only has a few classes during his last semester and he needs the money badly. He was considering borrowing the money to get him through the last semester. However, he has quite definitely made up his mind to accept employment with a firm he worked with the previous summer. This firm has just offered him a position upon graduation and he plans to accept the offer in the next week. Should he accept the part-time employment? He can delay the acceptance of the pending offer from the firm he worked for during the summer for a month and give the firm with the part-time work a trial period. Is it ethical to do so? If he does this and at the end of a month he decides he does not like the part-time firm and accepts work with the firm he previously worked for, should be then stop working part-time even though he needs the money?

Problem 5: A graduate student is working part-time for a local engineering firm. In his research work at the university he has access to the laboratories and laboratory equipment. He is approached one day by the head of the firm and asked to take six concrete cylinders back to the university with him and run compression tests on the cylinders. He is told to include the time he spends in the testing of the cylinders on his company timecard. He knows that the university does some commercial testing on a charge basis. He is quite certain he can test the cylinders without going through the university engineering department and he assumes the head of the firm might be pleased if he did so. What procedure should the student follow? What are the implications of the possible courses of action he can take?

Problem 6: A young graduate engineer is on his first job. He is assigned as an inspector on the construction of a large industrial building. At Christmas time he receives in the mail a nice tennis racket as a gift from the general contractor. The card wishes him "Merry Christmas" and thanks him for his help and cooperation. What course of action should the engineer follow?

Problem 7: An engineer is sent by his company to another city some distance away to evaluate a piece of test equipment his company is considering buying. They are also considering several other similar products by other manufacturers. While making this trip he has to stay overnight and he has taken his wife with him. Upon arriving at his hotel room he finds two tickets to the local concert of a nationally renowned symphony orchestra. He and his wife love music and have made no plans for the evening. The tickets are the courtesy of the manufacturing firm he is to visit. Should he use the tickets? If he does can he be entirely objective in his evaluation of the test equipment the next day?

Problem 8: An engineering firm is contracted to a utility company to make a feasibility study on the construction of a power plant in a given location. This firm has also designed many power plants and feels certain that if they determine that the plant is economically feasible they will also be given the contract to design the power facility. The firm is experiencing a slack period and a contract of this size is needed.

In the process of preparing the study it begins to appear that the power plant will not be economically feasible and there are a number of environmental problems with a location in the general area. By estimating somewhat low on the construction costs, the economics would be changed so that on paper the plant would be feasible. The environmental problems could be minimized. In essence the feasibility report could be slanted so that the utility company would likely go ahead with the design. Since many aspects of a feasibility report are estimates, the engineering firm feels justified in preparing their cost estimates so that the utility company will go ahead with the design and construction. Are there any ethical obligations upon the engineering firm in preparing a feasibility report? Should an engineering firm which is hired to make a feasibility study be considered for a followup design if the study shows the project to be sound? Should there be restrictions in ethical codes against the same engineering firm doing the design if they also had prepared the feasibility report?

Problem 9: Design engineering Firm A has been hired to do the design on a new highway. The resultant design is routed through a section of a city, and because of the resultant environmental impact there is considerable opposition to the highway location. The state highway department has approved the design and has declared it is the most logical location for the much needed highway.

A group of citizens and local organizations contact engineering Firm B and ask them to make an independent cost study of the highway location. They offer a reasonable fee to make this study and offer to increase it 25 percent if the firm is successful in determining another route which will be more economical. Is it ethical for Firm B to make this review and evaluation of Firm A's work? Is the 25 percent increase ethical?

Problem 10: Two engineers, after several years of experience working for a consulting firm, decide to form a partnership and start their own consulting firm. Upon mutual agreement one of the partners is to run the design office and the other is to make contacts with possible clients. As usual, work is slow at first and at an engineering meeting, a friend of the office partner indicates that he is temporarily overloaded with work and that he could use some help on some design work. He says he cannot pay a full consultant's fee but could pay him designer's wages on an hourly basis. The partner was told that it would be satisfactory to do the work at his office or at home in the evenings or weekends. The office partner had some heavy medical expenses at home and since the new venture was going very slowly he felt he could use the extra income very much.

The office partner is sure that if he told his business partner, his partner would feel the income from this work should go directly into the company account and be divided between them. He is also concerned that his partner would be angry that he would be taking the work on the low hourly rate instead of insisting on a consultant's fee, which would double the amount.

Should the office partner go ahead and take the work and disclose nothing to his partner and do the work at home in the evenings and weekends? Should he take the work and do it at the office during slack periods and when his partner is away? Should he do the work in the office and split the money with his partner even though he knows his partner spends company money taking prospective clients to lunch and sporting events?

Problem 11: A government employee has a responsible position with a government engineering bureau. The government bureau is to let an engineering design contract in the near future to one of three firms whichever appears to be most capable of performing the work. At a social gathering one night, the government employee is approached by the head of one of the firms and told that if his firm receives the contract, the government employee could come to work for him at a considerable increase in salary. The government employee has been thinking of leaving government service because the office location is a long commuting distance from his home. He could leave and withdraw his retirement pay and purchase a small orchard which he has been looking at for several months. The orchard is only a short distance from the offices of the consulting firm, located in a suburban area.

The government employee makes no commitment to the head of the consulting firm but keeps thinking about the advantages of the change in employment. A couple of weeks later he is in a meeting with the government engineers responsible for the decision on which firm to select for the design work. The choice is narrowed down to two firms, one of which is the firm that made the covert approach. No decision is reached that day. They are all told to study the two proposals and then reach a decision within three days. The government engineer is sure he can swing the decision to the firm which approached him by using mild persuasion on one or two members of the selection committee. He feels that both firms are capable of performing the design; in fact, it may be true that the firm from which he received the offer was in a slightly superior position.

He wonders if he should make a contact before the selection committee meets to see that the offer still holds and if so, to have a firm commitment on salary and position within the company and to be sure the company would wait six months so that no one would be suspicious.

What would you do if you were the government engineer? Is your considered course of action the proper situation ethics?

Problem 12: The owners of a medium size consulting engineering firm, XYZ, need work, and the authorities of a nearby county have authorized a new water treatment plant that will serve several communities. The county authorities believe that the best procedure in determining the design firm for the water treatment plant is to select the six most qualified engineering firms in the state and ask them to submit proposals giving their experience and qualifications along with a firm bid price for doing the design work. The invitation for proposals indicated that in all probability the contract for the design would go to the lowest bidder but that they reserved the right to reject any bidder they deemed unqualified. The specifications for the requirements for the design were very general, in that they required complete construction plans and specifications along with all necessary construction contract documents. The quality of water together with volume output of the plant was specified.

Firm XYZ was sent an invitation to bid. The owners of this engineering organization realize that they must obtain this work in order to retain their employees. They know that the competition will be keen, since engineering work has been slow in the state. They reason that they can minimize the engineering time because one of the principals has a complete set of drawings and specifications of a similar size water treatment plant he designed about ten years ago while working with another firm in a distant state. Firm XYZ considers that may parts of the drawings can be traced, so they submit a low bid—about 70 percent of what would normally be required to do a design job of this magnitude.

Firm XYZ is the lowest bidder by a considerable amount of money, which makes them disgusted that they didn't bid somewhat higher. Since they bid so low, they decide they must make further savings in design cost or there will be no profits. Many of the design calculations are simplified with assumptions on the high side so that the facilities would be overdesigned and not underdesigned. Firm XYZ considered they were justified in their action since they were doing the job so cheaply. Is this so? Were there any breaches of a code of ethics?

Problem 13: One of the engineering firms which bid on the design work in Problem 12 was perplexed to see how firm XYZ could bid so low for the work. When firm XYZ finished the job the county advertised for construction bids and made sets of plans and specifications available for a nominal price. This second firm purchased a set and made a study of the design. They determined that the design called for equipment that was not up-to-date and that several pumps and motors were oversized. They estimated that a rigorous design could reduce the cost of the plant by more than a reasonable design fee. Also, more modern and efficient equipment should be specified. The firm of engineers were uncertain as to the course of action to take. The alternatives were:

1. Forget the whole thing and let the county go ahead and build the poorly designed plant.
2. Go to the XYZ firm and demand they redesign the plant.
3. Tell the story to the county commissioners and show them what it was costing them for a cheap design.
4. Give the story to the local news media and demand an exposé.

What is the best course of action? Are there any other alternatives than the above four?

REFERENCES

1. B. J. Lewis, "The Story Behind the Recent National Scandals Involving Engineers," *Engineering Issues*, ASCE, April 1977, pp. 91–98.

5. ENGINEERING
ORGANIZATIONS

Since early man took on the role as leader and director of construction projects, he has usually fitted into an organization of some form. Today there are a multiplicity of engineering organizations that provide the environment for the working engineer. Such organizations may vary from a one-man office to a corporation employing thousands of engineers.

Generally speaking, organizations employing engineers can be divided into three broad groupings: consulting firms, industrial organizations, and governmental bodies. The only other category that might be listed separately would be purely research firms. However, they could reasonably be classed as industrial organizations. These three basic organizations will be described here, and the role of the engineer in each will be given. Since the activities of engineers are numerous, it will only be possible to describe the work in broad terms. However, the engineering student should have a general picture of his future area of activity. He will then be able to better understand contracts and the elements of contract law which will follow in subsequent chapters.

A. CONSULTING FIRMS

An engineering consulting firm is a business entity that is formed for the purpose of providing engineering services for gain. Such engineering services are usually an activity that plans, evaluates, and designs a project. Projects may be of a multitude of kinds, from a real estate development, bridge, or industrial plant to a biomedical device. Feasibility reports, including economic studies of a proposed project, may also be the product of a consulting firm.

An activity found in most consulting firms is the preparation of plans and specifications in a suitable form so that contractors can prepare bids and then execute the work. Consulting firms rarely perform the construction work, although a consulting firm may have resident engineers and inspectors at the construction site to see that the construction is performed in accordance with the plans and specifications. However, there are business organizations that will prepare designs and also perform the construction for a fee. More details of such operations are given in Chapter 9.

Consulting firms start with an engineer or a group of engineers who, after judging that they have the necessary experience and expertise, form their own organization. Such a business organization can be in the form of a partnership or a corporation. Just a few decades ago, most consulting firms were formed as partnerships, but in recent times corporate organizations appear to have advantages. This is especially true as the organization grows in size.

There are several advantages to a partnership and also several disadvantages. A brief summary is given here.

Advantages

1. Partnerships are usually formed with two to four members, although they can consist of more partners. The legal rules of formation are simple. A partnership can become a legal entity by a simple statement signed by the partners setting forth the percentage ownership of each partner in the organization. The responsibilities of each partner should be given and any limitation of action of any partner should be so stated in the partnership agreement. Partnership agreements do not have to be approved by any government bureau. Such agreements spell out the way in which profits and losses are shared. Simplicity of the process of organization is a great advantage.

2. In a partnership, one partner cannot sell his portion of the company without the consent of all the other partners. In this manner, the partners have close control over the management and ownership.

3. Generally speaking, a partnership is not subject to income taxes. All profits are taxed on the basis of individual income taxes to each of the partners. This usually amounts to lower taxes, unless an individual has such a large income that he is in the upper tax brackets.

Disadvantages

1. A partnership is limited in capital to those monies which the partners themselves can raise. Of course, a partnership can borrow money, but they cannot sell interest in the company to a nonpartner.

2. Partners are individually liable for the debts of a partnership, and assets that are individually owned will be subject to execution to satisfy any such debt when partnership assets are insufficient.

As previously stated, the present policy of most consulting firms is to incorporate. A corporation is an association of shareholders created under the law and regarded by the courts as an artificial person. It is a legal entity separate and distinct from the individuals who, as shareholders, comprise the corporation.

Corporations are formed under rules of law established by state legislatures. A corporation cannot be formed without approval by the Securities and Exchange Commission of the state, and cannot be dissolved without approval by the same Commission. This is to protect the public from the fraudulent sale of stock and abuse of power by the officers. Since the formation of a corporation must conform to the law, a lawyer specializing in corporate law should be retained to prepare the necessary papers of incorporation.

A corporation can hold property, sue and be sued, and exercise other powers bestowed upon it by law. Corporations are owned by the stockholders, who in turn elect the Board of Directors, who in turn appoint the management. Stockholders are not personally liable for the debts of the company. Also, they only have a vote equal to the number of shares which they own. One who owns only a small percentage of the total stock has little voice in the management.

Engineering consulting firms which incorporate usually do so with a limited number of shareholders. This is spoken of as a closed corporation. Written agreements may be part of the by-laws, so that if a person wishes to sell his stock and withdraw, he must first offer it to the other stockholders of the firm. In consulting firms, mostly all stockholders are working members of the company.

A common provision in the by-laws of consulting firms is that at the death of one of the principals, the stock is to be sold to the surviving principals. By this provision, the stock remains among the organizers of the company and will not fall into the hands of nonengineers.

Other advantages of a corporation are as follows: (1) Stockholders are not individually liable for the debts of the company. In today's world of liability claims, an engineer can protect his personal assets by forming a consulting firm as a corporation. (2) Capital is acquired by selling stock and, if desired, capital can be raised by selling stock to persons other than those involved in the management.

There are other disadvantages of a stock company besides the cost and involvement of the organization. Corporations are subject to federal and state income taxes. These taxes are based upon net profits before any payment of dividends. As of 1979, the federal tax rate for corporations is 17 percent for the first $25,000 of profits; 20 percent for the second $25,000; 30 percent for the third $25,000; 40 percent for the fourth $25,000; and 46 percent for profits over $100,000. The total income tax comparison between a partnership or a corporation would depend upon total net income of the individual.

It can be much more difficult for a major stockholder of a corporation to withdraw his participation in the company than it would be for a principal in the partnership. A partnership can be dissolved upon the withdrawal of any one member. However, in a corporation a principal member must sell his stock in order to withdraw. He cannot dissolve the corporation unilaterally unless he owns more than fifty percent of the stock.

Consulting firms seek engineering contracts from many sources. Civil engineering firms may find the bulk of their business with government organizations such as city, county, state, and federal bureaus. Consulting firms specializing in structural engineering are most likely to have contracts with architectural firms who are the principal contractors for the design of a building. Electrical and mechanical consulting engineers will also have parallel contracts on the same building for the preparation of plans and specifications for their phase of the work.

The engineer engaged in consulting engineering will be the middle-man in the project triumvirate. This group will consist of the owner-to-be of the project, the engineer consultant, and the contractor who will build the facility. The owner may have the design engineer monitor the work of the contractor, or, as in the case of governmental agencies, the owner may do that himself. Since the consulting engineer is heavily involved in contracts between himself and the owner, and also in the preparation of materials for the contract between the owner and the builder/contractor, it is very necessary that he have a good understanding of contract law.

There are a number of firms who do both the design and construction of engineered facilities. These are generally large organizations that direct their work to industrial plants, energy facilities, mining and smelting operations, etc. They generally work in the private sector and not in government supported work. In the latter case, the design and construction phases are almost always separate.

This combined design and construction procedure carries the name of "turnkey operation." It usually has the advantage of great speed in getting the facility built and may result in fewer legal complications, especially where changes are desired by the owner after construction has started. The turnkey method usually works to the advantage of the owner when the facility is very complex and must involve close cooperation between those doing the designing and the building. In some instances of turnkey operations, the construction has started before the final design is complete.

The disadvantage of the turnkey operation is in the lack of competition in construction costs. The contract between the owner and the turnkey firm calls for the payment of costs by the owner plus a fixed or percentage fee to the designer-builder (turnkey). This uncertainty of final costs at the beginning of the project may be tolerable in the private sector, but is usually not tolerable in the public sector.

B. INDUSTRIAL ORGANIZATIONS

1. Manufacturing

Many engineers are employed by industrial organizations in a wide variety of activities ranging from research, design, operation, and maintenance to sales. A higher percentage of chemical, electrical, and mechanical engineers will have such employment than will civil engineers. This is especially the case in strictly manufacturing organizations. A bachelor's degree is usually sufficient, except for research and design work. For research work, a doctor's degree is the desired training, while a master's degree is the preferable academic preparation for design.

Industrial organizations will have lawyers within the organization or have ready access to legal firms for their activities involving contracts. However, top management must themselves have an understanding of the legal aspects of contracts, liability, and employee relations.

Some manufacturing companies may rely heavily on a large engineering staff, while others, due to the specific nature of the business, may employ few or no engineers. In recent years, a larger percentage of top management have had academic training in engineering.

2. Construction

Construction has been big business in the United States ever since the depression of the 1930s. Even before that time, the United States was noted for its big construction projects, such as the system of railroads, the Panama Canal, the Hoover and Grand Coulee Dams, large suspension bridges, and the Empire State Building, to mention a few. In past years, the managers of construction firms were most likely to have risen from the ranks of the building trades or to have been part of a family ownership and thus developed their knowledge of construction from on-the-job training. Today, this trend is giving way to the management of construction by graduate engineers.

The complexities of present-day construction call for knowledge in various areas, including engineering, business principles, contract law, and labor relations. In recent years, several academic programs have been developed in construction management at engineering colleges. These programs usually terminate in a master's degree.

The construction firm works from drawings and specifications prepared by the design engineer. His work is inspected for the purpose of ensuring that the work conforms to the contract plans and specifications. In most instances, the contractor has the freedom of performing the work in any manner he desires, but the final result must conform to the design plans and the materials must be those specified.

The details of contract relationships between owner, design engineer, and contractor will be given in Chapters 8 and 9.

3. Research and Development

Research and development (R&D) organizations employing engineers can be part of an industrial organization or they can be an entity in themselves. The number of engineers employed in R&D is a small percentage of the total engineering work force. This type of work requires a high degree of technical knowledge in a relatively narrow field of activity. As such, the technical employee usually must have a graduate degree.

A research organization employing engineers might be a private institution or part of an industrial organization, or might belong to a government agency. Funds for operations may come from the profits of the organization, from part of an industrial firm's budget, or from tax revenues. Many large industrial firms owe their existence to new products developed through research. The field of chemical engineering has the highest percentage of engineers employed in research and development. This is followed by electrical engineering, mechanical engineering, and last by civil engineering.

C. GOVERNMENT ENGINEERING

The role of government in engineering is quite complex and covers a wide range of activity. One of the first activities of the so-called "engineer" was in military operations. The word "engineer" comes from the latin "ingenear," meaning one who is ingenious. An ingenious person was one who developed and built machines of war (catapults, etc.). He was a person apart from the warrior. In more complex military operations, he also designed and supervised the building of breastworks, fortifications, and other defensive works. He had the title of military engineer. Leonardo da Vinci advertised himself as such a person. The activity of construction was developing about the same time period. It was not related to military engineering and became known as civil engineering. Most of the early civil works, such as roads and streets, water supplies, canals, etc. were sponsored and paid for by government. However, work along these lines was also done by private companies. The early turnpikes in the Eastern United States were built and operated by private companies.

1. Cities

City governments of medium to large size have engineering departments. These may vary from a one-man department to a team of several hundred in a large city like Los Angeles or New York City. The duties of an engineer in a city department may be varied. The main functions of a city engineering department are to plan and maintain roads, streets, and utility systems, and to review plans and specifications for new buildings, either public or private. The monitoring of zoning ordinances may be an activity of the city engineering department and also the inspection of building construction for compliance with the city building code.

Larger cities may have sufficient staff to prepare designs and specifications for new streets, storm drainage systems, and utility extension or renovation, as well as new city-owned buildings. It is usually not economical for a small to middle-sized city to maintain an engineering staff for these activities since the volume of work in any one of the specialty areas is infrequent. The usual procedure is to contract such work to consulting engineering firms, with the engineers employed directly by the city supplying input data to the consulting engineer and also reviewing the plans.

Except for the case of employment on a large city engineering staff, the engineer in the service of a city will have to be knowledgeable in several diverse specialties of engineering. The training for such work is usually civil engineering, although electrical engineers are needed in the electrical utility department.

Very large cities may have special bureaus established by law that have bonding power to operate somewhat independently of the city government and of the city engineering department. The function of such organizations consists of the design, operation, and maintenance of specific areas of public works. Examples of such entities are the New York Triborough Bridge and Tunnel Authority, and the Los Angeles Water and Power Board. These engineering organizations employ civil, electrical, and mechanical engineers.

Even though a city engineering department is large, they may still contract design work to consulting engineers for special projects such as large bridges, buildings, or power stations.

2. Counties

The county government units in the United States, in general, have jurisdiction over rural rather than suburban areas. There are some exceptions to this, however. When the county is nonurban, the demand for engineers is not great; therefore, such counties have very small engineering organizations. Responsibilities of county engineers include roads and bridges, storm drainage facilities, subdivision approval, and issuing of building permits within the county. Design of new facilities of medium to large size will be performed by consulting firms under contract to the county.

Because the engineering work in the counties during the early history of the United States was primarily surveying, the person responsible for such work carried the title of "County Surveyor." This holdover title is still prevalent in many counties, although the duties may now involve more engineering than surveying.

3. Federal Government

The federal government of the United States has been involved in engineering since its beginning. However, engineering work in the 18th and 19th centuries was very limited. As previously mentioned, the early roads were built by local governments in a limited geographical region, or the longer turnpikes were constructed and operated as toll roads by private companies. The building of canals was usually financed by private funds and operated on a profit concept, though not always successfully.

The federal highway system did not begin until 1916 when the federal government authorized the expenditure of 75 million dollars for rural road improvement. However, it was not until the Federal-Aid Highway Act of 1921 that designated interstate highways were established.

The first major involvement of the federal government in large construction was the building of the Panama Canal in the early part of this century. The U.S. Army Corps of Engineers was primarily a military organization. However, they had been involved in public works such as flood control, navigation, dredging, etc. In this work, the Corps had hired civilian engineers, since the military did not have sufficient educated and trained engineers.

When work on the Panama Canal faltered under private management, the United States government assumed control and turned the responsibility of the construction of the canal to the Corps of Engineers. The success of this operation was a shot in the arm to the civil branch of the Corps, and this federal engineering organization became and still is responsible for flood control and maintenance of navigation waterways throughout the country. It has military personnel for top management, but detailed engineering is performed by civilian engineers working in civil service status. The bulk of the engineers of the Corps have their training and responsibilities in civil engineering, although mechanical and electrical engineers are also present in the various offices throughout the United States.

The Corps of Engineers contract the major design projects to private consulting firms and the construction is done by private contractors under contract to the Corps. Engineering personnel of the Corps are extensively involved in contract preparation, management, and inspection.

A second federal government organization that employs many engineers is the Bureau of Reclamation. Their major charge is to provide irrigation water for the arid regions of the western states. In this role, they have been responsible for many dams, canals, irrigation systems and new roads and bridges where new reservoirs have required such. Although the Bureau does a high percentage of the required design work with in-house personnel, it does at times contract for design work and does contract for all construction work. Some of the largest nonmilitary construction contracts ever let have been the large dams in the West.

The federal government employs many engineers in many other organizations such as the Federal Highway Administration, EPA, and NASA, just to mention a few. The military services have many engineers on their payrolls, both in military personnel and in civil service status.

A large percentage of government engineers are involved in contracts to private organizations for consulting, research, and procurement of supplies and facilities. Almost all contracts are reviewed by government employees with legal training, but the writing of contracts and the preparation of specifications can be the duty of the engineer. Therefore, an understanding of contracts and their ramifications is necessary for many engineers employed by the federal government.

D. ENGINEERING REGISTRATION

In order to protect the public from imposters, all states have laws that require people who work in the private sector in professional occupations (such as engineers, doctors, and lawyers) to be licensed by the state. Such licenses require the passing of an examination. The professional engineer examination consists of two parts of one day's duration each. The first-day exam covers engineering theory and is broad in content. It is the same for all branches of engineering. It is called the Engineer-in-Training Exam and can be taken at the end of the training for the bachelor's degree. It is preferable for the future engineer to take this exam at that time. There is a national uniform exam available for the E.I.T. and most states give this exam. The second-day exam can only be taken after meeting the experience requirement. This exam is in one's branch of engineering, with optional questions so that a candidate can have a majority, but usually not all, of his problems in his specialty within his branch of engineering. The second day's exam covers the design aspect of engineering. For an engineering license, a period of work experience is required after the obtaining of a baccalaureate degree in engineering from a program of instruction accredited by the Engineering Council for Professional Development (now the ABET). In most states, the minimum required work experience is four years in the area of engineering in which one wishes to be licensed.

The evaluation of work experience as well as preparation and grading of the examinations is performed by a Board of Engineers appointed by the governor of the state. The major engineering societies within the respective states usually make recommendations to the governor for the filling of positions on the registration board. A recent trend is to have some members sit on the board from outside the engineering profession. States such as California have passed such laws, and nonprofessionals sit on professional registration boards. The thought behind such laws is to see that registration laws and actions of registration boards do not become unduly restrictive in limiting the number of registered professionals. A weakness in this reasoning is that a nonprofessional will have difficulty in evaluating work experience, the severity of the examination, and also the severity in the grading of the examination. About all a nonprofessional can accomplish in a role as a board member is to prevent any overt action by a board. Such restrictive activity by engineering registration boards has not been proven. The rare cases of complaint have usually come from unqualified people seeking an easy way into the professional ranks.

Unqualified people practicing in the highly technical field of engineering can pose a threat to the life and limb of many people as well as property loss of hundreds of thousands of dollars.

There are provisions in the registration laws that a license can be withdrawn when an engineer is proven to be incompetent. This can occur—and has occurred—where failures have proven to be the result of poor or inadequate engineering. Revoking of a professional license is also possible when "unethical behavior" is proven. In most cases, unethical behavior is construed to be the breaking of the law of the land. The registration laws do not contain codes of ethics, so only the conviction of a felony or incompetent work can be the grounds for loss of license.

Not all engineers are required to be licensed. In fact, a minority of about one-third of the total engineers with the requisite experience take the examination and become registered. The highest percentage of registration is among the civil engineers, with considerably lower percentages in the branches of electrical, mechanical, and chemical engineering.

Each state has a law requiring the person who is responsible for the preparation of plans and specifications of engineering works to be used by the public be licensed in the state where the facility will be built.

This, then, requires that of all the people who perform work on the design, only the persons putting their signatures on the drawing need to be registered. The person or persons are then the ones responsible for the project. This will usually be the chief engineer and/or the project engineer.

Engineers who work for government agencies are not required to be registered as long as the work they do is in the service of the government. However, many government agencies will give preference to potential employees if they are registered and have the necessary experience. This is especially the case where a city is employing a person as the *City Engineer*. Many cities by law will require that the person with the title of City Engineer be registered. Counties may also require registration for the position of County Engineer. If lawmakers see the need of private engineers to prove their qualifications by meeting the professional registration requirements, it seems reasonable that public engineers who supervise the spending of thousands of tax dollars should also be required to meet the requirement of professional registration. The author was astonished many years ago when he submitted a subdivision plan to a City Engineer for approval and upon obtaining the approved plans was told by the City Engineer that he must be a very excellent surveyor since the boundaries of the subdivision and all lots closed. The City Engineer's background had been limited to a survey crew; yet he did not know that the descriptions of all land surveys must close and the field work must conform to the lot description.

The engineering profession should be active in educating the lawmakers to the need for professional registration of all government employees carrying the title of Engineer.

All consulting firms will have their principals and also their project engineers registered. A progressive firm will aid and encourage all employees responsible for engineering to obtain the necessary training and experience to qualify for registration. Engineers working in industry, such as manufacturing firms, usually see little need for registration and therefore do not become registered professional engineers. Some may do so, however, since it does indicate a degree of competence and professionalism.

The American Society of Civil Engineers requires professional registration for the highest grade of membership, which is designated as "Fellow."

A great threat to the public welfare is the activity of those who offer their engineering services to the public, but lack competency for registration. State registration boards in many cases do not have the authority to take disciplinary action against nonregistered engineers and their work.

District attorneys are too busy on major criminal offenses, and will drag their feet against prosecution of those who violate the registration laws. Such violation is most often by those who are not competent enough to perform engineering design. Only in the case of failure of an engineering work in which there is loss of life, injury, or heavy property loss will action likely be brought against nonregistrants.

Twenty-one state registation boards reported on their law enforcement activities during 1978 for the National Council of Engineering Examiners. California listed five cases where civil engineers or land surveyors had registration revoked or suspended for reasons ranging from practice that was beyond the scope of civil engineering to negligence and incompetence. In Maryland, thirteen registered engineers were found guilty by a federal court of being involved in a kickback arrangement. In Texas there was a case of forging an engineer's seal by use of an electrostatic copier. Two engineers in New Mexico embezzled $395,000 from the state. In Oregon an engineer lost his license because a reservoir he designed collapsed.

From the above sampling, it is seen that registration boards are willing to look into misconduct cases of registered engineers and take action where necessary. There are increasing pressures upon registration boards from engineers and the public to remove from practice incompetents and those who break the law. Enforcement against those practicing engineering without a license should also be active.

In Iowa the Supreme Court ruled that a professional engineer's license could be revoked by the State Board of Engineering Examiners upon showing of misconduct, and that it was not necessary to show gross misconduct, as the defendant claimed. The Board fround that the defendant had misstated the design snow load in the design of roof trusses. The Board maintained that the plans were intentionally misleading and were a betrayal of public trust. The state law gave the grounds for revocation as "any gross negligence, incompetence, or misconduct." The courts ruled that the word gross was an adjective modifying negligence and not incompetence or misconduct. Here is a case where the ruling on the wording of a law is given.

New developments have come out of a February 1979 ruling by the U.S. Supreme Court. This ruling was that states may bar members of a profession from practicing under a trade name. Texas and 19 other states have such a law. The 1969 Texas law was challenged by a firm of optometrists practicing under a trade name and the U.S. District Court ruled against the state law. This district court ruling was overturned by the Supreme Court. The higher court stated that the possibilities for deception would be very numerous if the Texas law was not upheld. The Texas Optometry Association has an ethical ban against using trade names.

In the same ruling the Supreme Court upheld the Texas law that requires two-thirds of the members of the state licensing board must belong to the Texas Optometry Association. It also rejected a proposition that the general public be represented on the board. This contrasts with California, where the majority of the members of the state engineering board are "public members."

6. LAW

A. INTRODUCTION

Law is a system of rules which govern human action. Society has established those rules in order for people to live in harmony with one another. Since man is fallible, laws and enforcement of laws are also fallible, and their harmony is relative. Laws are regulative. They regulate interaction between the government and the individual and also the relations among individuals. Laws require the relinquishing of certain rights for the good of society. A simple example of this are the laws that govern the operation of an automobile. Consider the end result if tomorrow all laws were abolished relating to the driving of a motor vehicle.

Laws can be divided into three categories: natural laws, spiritual laws, and man-made laws. Spiritual laws are those which are accepted as having come from a divine power, whether directly from revelation or through human mouthpieces. Aquinas wrote [1], "man is ordained to the end of eternal happiness, and since salvation is a supernatural end which exceeds man's power to achieve without God's help, it was necessary that man be directed to this end by a law given of God." Such spiritual laws have had a definite influence on the development of man-made laws. This will be discussed further in a later section of this chapter.

Another set of laws which control man are the natural laws which are part of the earth as well as the universe. Man has had to live in conformance to these laws, and through the ages has been trying to understand and use these natural laws to his benefit. The 20th century has seen phenomenal growth in the understanding and use of the physical laws of the universe to supply man with inspirational and material needs. As man has learned, and sometimes the hard way, that he must understand natural law in order to work within it and use it for the well-being of mankind, some thinkers also believe that there is an absolute set of moral laws that man must conform to in order to have a joyful life. Society has, since the beginning of time, tried to fashion laws of human conduct. Some of these laws are so basic to the safety and happiness of the human race that they have been part of the written laws of society for ages. Other laws become obsolete and new laws take their place.

An important difference between natural physical laws and the laws of human conduct is that the former applies to all things while the latter are addressed to man alone. The natural laws require a necessary behavior while the laws of human conduct imply freedom to obey or not to obey. However, refusal to obey either sets of laws will exact a penalty. In natural law, there is no way to escape the penalty of disobedience. The breaking of man's laws is nearly always under a unique set of circumstances and judgment is passed by fallible man, so that justice is not always given.

Law is a director of human activity. It can direct the road that a country or a group of people within a country will travel. Laws relating to freedom of religion, speech, and action determine the politics of a country and whether there is peace on earth or constant turmoil among nations.

The first legal profession was developed in Rome. The Law of the Twelve Tables was drawn up in 451-450 B.C. These Tables were based on the Roman religion and they set forth customary manner of behavior. The laws were inscribed on brass tablets. The Romans had judges called proctors who in addition to judging disobedience against the Tables also made decisions that had the effect of law. In addition to the Law of the Tables, which applied only to Roman citizens, the Romans later developed a system of law called the law of nations. These laws were an attempt at developing a system of laws on the moral instincts of all civilized men. Under the Emperor Justinian the laws of Rome were classified (codificiation). These codified laws had a great effect on later European law. In the Middle Ages the new national states of continental Europe adopted the civil law of Rome.

B. WRITTEN LAW

Written law is a set of rules established by legal bodies for the control of human conduct. These laws are sometimes referred to as positive law. Though moral law was, and still is, a strong influence on the development of positive law, it is distinct and separate from positive law. Positive law defines and enforces the rights and duties of all people within the jurisdiction of the written law.

Business transactions are controlled by law. Without such laws business could not operate. Law has defined the duties of all parties to business contracts. The industrial and commercial world of modern countries is very complex. Large amounts of money are risked in commercial enterprises. Risk can be minimized by laws that give investors some assurance that all parties will operate within rules that require honest and predictable action. Countries which have lacked sufficient laws and enforcement of such to give confidence to investors have also lacked commercial and industrial development. Without knowing the duties and rights under the law, one will be at a considerable disadvantage when operating in any business activity. Since engineers work on projects that involve moderate to large amounts of money, it is necessary that they have a fundamental understanding of the laws that legally control their course of action in relationships with other people or organizations. The law says that ignorance of the law is no excuse. In a complex society such as the one found in the United States, fundamental knowledge of the laws pertaining to one's field of activity is very necessary. However, in addition, it is also necessary to have people available who can advise on more complex and detailed matters of the law. A large percentage of the activity of the legal profession is the giving of advice and the preparation of legal documents.

Written law in the United States is found in two categories: (1) the Constitution of the United States and the Constitutions of the fifty states; and (2) the acts of Congress, the state legislatures, and municipal legislative bodies. The Constitution of the United States is the supreme law of the land. No law, or ruling of law, nor individual state constitutions can be in violation of or contrary to the United States Constitution. The purpose of the U.S. Supreme Court is to judge and rule on whether any written law or ruling of law is contrary to the Constitution. The individual state Supreme Courts are for the same purpose. Constitutions consist of general fundamental principles.

Statutory law is enacted by elected legislative bodies and relate to specific acts or relationships of people and organizations. Legislatures through the centuries have passed laws that have in effect established what is right and wrong and also the possible punishment of disobeying the law. These laws have covered civil acts as well as criminal. The law that engineers deal with in the operation of their profession is civil law.

Breaking of a criminal law in engineering matters is indeed rare and is restricted almost to fraud, perjury, or willful negligence.

There are other sources of law that have come to play an important role in the making of laws. These sources are administrative agencies at the federal, state, and local levels. In recent years the federal government has passed a number of laws creating regulatory agencies. These agencies have been given the mandate to regulate and enforce some activity of the people (including business and industry) that Congress has decided was not operating in a manner that was protecting the well-being of society in general. The EPA and OSHA are two agencies on the national level which have interacted with the engineering profession on a national level. Congress did not write the regulations under which a specific agency was to act. The agencies were given the power to write and enforce the needed regulations. This has created another group of governing, law-writing bodies—in effect, lawmakers which have not been elected by the people.

On the local level licensing boards and zoning commissions also create laws on a lesser scale. Such regulatory agencies, however, do have closer control than those in the federal system.

Civil Law. Civil law is derived from specific acts of legislation by a political power consisting (in the United States) of persons elected by the people to act in a representative assembly for the people. A law can be classed as a civil law if it controls the action between persons or organizations as opposed to the acts against the state. Acts against the state are in the majority in the classification of criminal law. Civil law is often referred to as private law. This area of private or civil law deals with the rules relating to contracts, real estate, personal injuries, taxation, and business affairs. Most lawyers and courts spend the majority of their time dealing with problems of civil law. The Code of Hammurabi is the earliest known document containing codified civil laws.

Criminal Law. This part of law is most familiar to people, since the activities of criminal law are what usually appear in the newspapers and on the television screen. A criminal act is the disobeying of a written statute which all people within the state are required to obey. Most criminal law statutes are written by state legislatures and municipal authorities. There are federal criminal laws involving taxation, the importing of contraband material, and crimes that are interstate in nature. Congress can in effect make a federal law covering any damaging act. For example, in the 1930s kidnapping became a federal offense. The main purpose was to call upon the F.B.I. to assist in solving this form of criminal action, which had become prevalent.

Common Law. The words *common law* are sometimes used to describe the laws of English-speaking countries. England is the source of common law and the British settlers in North America brought this with them. Common law is what both American and Canadian law are founded on. Texas and California have traces of Spanish law and Louisiana still retains some of the Napoleonic Code. The law of Quebec is influenced strongly by the past French civil law.

In legal terms common law is referred to as the unwritten law that receives its authority from universal acceptance. It is distinguished from statute law which clearly defines illegal acts as declared by the will of the Legislature. The common law has its origins in court decisions that may be recent or may go back many years. Early America accepted those court decisions that referred to English common laws of reason. The term *common* comes from England because judicial rulings were applied uniformly to all parts of the land.

Common law has its origin in court decisions and is not founded on legislative statutes. The basic concept is that all interactions between people or organizations that may result in dispute cannot possibly be covered by statute. Therefore, when such disputes arise the courts will use good judgment and fairness in settling the dispute between the contending parties. Such judgments are then legally binding upon the contesting parties.

In early England, the ruler was the lawgiver and fountain of all justice. William the Conqueror and his successors asserted this doctrine and practiced this role of judge and lawgiver. As populations grew and business developed, courts were established to settle disputes among the people. These courts became common-law courts. Judgments and rules formulated by these courts then became the basis for judging future cases of a similar nature. Most of the states, by legislative enactment, have declared the English common-law system the basis of judicial decision in all cases which have not been covered by statutory or written law. As circumstances develop, there may be pressure by the people to change a rule of common law. This must be accomplished by legislative enactment.

The decisions by the courts of common law form judicial precedents which bear considerable weight on later cases within the same court jurisdictions. Court jurisdictions are usually within state boundaries. It has been known that similar cases have been judged differently in different states. When a decision is rendered in a court of common law, it acquires a name and citation which identifies the case for future references. All such cases are recorded in official legal publications. Law libraries preserve these case histories and are very important in the training of lawyers, as well as a source of research when future cases develop. A skillful lawyer must be adept in researching future cases. A recorded judicial decision should give the pertinent facts of the case, the decision arrived at, and the applicable law or laws. The published court decisions are very vast, and grow larger each year. A law library must have trained librarians who can assist those needing past histories of cases similar to the case in question.

A development in English law that has had and does have a great importance in American law is the principle of *equity jurisprudence.* The early English court system was very rigid and limited in the jurisdiction in which it would operate. The limited scope of action of the court system caused petitions to be brought to the king in order to settle those disputes that were outside the law. When petitions became numerous, the king would then assign most of the cases to his chancellor. The English Chancery Court was thus developed which was later called the Court of Equity. This latter name came about because justice was dispensed not on the basis of any written rules or laws, but on what was fair and equitable. Abstract principles of justice were absent in such courts. The decisions rendered by the Chancery Courts were recorded, and successive chancellors considered themselves bound by the prior decisions of the courts. From these recorded court decisions a body of law grew that was in addition to statute law. Today the same principles of equity law, extensive in content, comprise the body of common law.

There are general rules and maxims that govern decisions in equity jurisprudence. They are:

1. Acts must be against persons and not just against the world.
2. Equity will not suffer a wrong without a remedy.
3. Equity has power to act against the person. It can command him to do what needs to be done or restrain him from doing an act which has been prohibited by court decree. The latter form of prohibition is termed an injunction.
4. He who seeks equity from another must be blameless.

There are other equity principles that are recognized by the courts. Separate chancery courts exist in about six states, but in the other states, the regular courts of law hear the cases of equity, and now there is only one set of courts. An equity judgment can order (if physically possible) a delinquent party to complete a contract.

C. LAW ENFORCEMENT

A person can only break a written statute. If he fails to perform his part of a contract, he is not breaking the law. The kinds of legal cases wherein the law is not broken fall into the category of common law or equity jurisprudence. Enforcement of the written law is generally the responsibility of the government unit that wrote the law. However, today, arrests can usually be made by any law enforcement agency irrespective of what division of government wrote the law that was broken. For example, a criminal suspect wanted in one state for robbery can be arrested in another state by the police of the second state. There are prescribed court procedures that must be followed, however, to have the suspect moved back to the first state to stand trial. The police of the state where the crime was committed can go into the second state and return the prisoner after a court order in the second state has been issued for the prisoners removal. However, the police of the state where the crime was committed cannot pursue the suspect across state lines and make the arrest in another state. This same principle applies to police officers on a city and county basis. Police of City A cannot make arrests in City B. In some localities city police may not make arrests in county areas outside the city. However, this has recently been changed for greater crime control by making city police deputy county sheriffs.

The law requires the police to arrest those who are in violation of the legal statutes. A private citizen can also make a *citizen's arrest* if he sees that a law is being broken. There are some types of laws in which the police do not actively seek lawbreakers and make arrests only upon being directed to do so by a person responsible for prosecuting such cases. An example of this is zoning ordinances. Action is usually taken by the constituted authority upon complaint by the citizens. In most civil action cases complaints have to be filed before legal action can be taken. It is not practical to have a police force of sufficient size to police such laws as zoning ordinances, etc.

Citizens have a responsibility to see that those who break the law are brought to justice. However, this must be done in a legal manner. Vigilante groups may have been necessary for law and order in the 19th century during the settlement of the West, but they have no place in the 20th century. The rise in crime in areas of large cities has been due in part to the apathy of the residents in these areas. Law enforcement needs the support of all citizens. Knowledge of the manner in which the judicial system operates can be of great help in matters pertaining to law and law enforcement.

Laws make courts and justices indispensable. Alexander Hamilton said, "Laws are a dead letter without courts to expound and define their true meaning and operation." Justices have the responsibility of deciding if the facts of the case bring the case under the specific provisions of the law. This is also the responsibility of the prosecuting attorney. The administration of the law is called *justice*. In theory, justice is the fair and correct administration of the law.

D. COURTS

Each country has its own unique judicial system, and the judicial system in each state of the United States is independent, except for the federal courts. There are several court divisions within each state. There is such a wide variety of court cases that a system of courts is necessary. Some cases involve matters that occur entirely within the bounds of the state and relate to state or municipal laws. Other cases involve infractions against federal laws or concern action that took place in several states. Therefore, there are two separate court systems; the federal court system and the state court system.

Federal courts operate within each state. The basic trial court in the federal system is called the *district court*. Each state has at least one federal district court within its boundaries. The more populous states have several district courts, each presided over by a federal district court judge appointed by the President of the United States and confirmed by the U.S. Senate.

Federal judges are appointed for life or until voluntary retirement. There have been federal judges who have served until old age. Some have remained capable in spite of advanced age, but others have become a problem to the federal court system.

If a party in a district court case considers that the judgment was not correct or fair, he may appeal his case to a federal *circuit court of appeals.* There are eleven separate circuit courts of appeals in the United States. In 1978 the total number of federal court judges was increased by nearly twenty-five percent to take care of the increase in the number of federal court cases.

Upon losing a case in the circuit court of appeals, the party has one last resort. This is to the United States Supreme Court. It is not automatic that an appeal to the Supreme Court will be presented before the Court. The Supreme Court reviews the petitions for hearings before the court and unless there is merit in the appeal it will be rejected. Cases that are important to the country as a whole are considered and cases questioning the constitutionality of bills passed by Congress gain a hearing before the Supreme Court.

There are several special federal courts which deal with unique types of cases. The Court of Claims hears and decides cases involving claims against the United States Government; the Patent Court decides disputed patent claims; and the Tax Court hears disputes between the government and taxpayers.

What types of cases are tried in the federal courts? All disputes involving the federal constitution and federal statutes are tried in the federal courts. A recent celebrated case that went through the federal courts and was finally settled at the Supreme Court was the case of Allan Bakke, who claimed he had suffered discrimination (loss of constitutional rights) when applying to medical school. The medical school had rejected Bakke, who was white, in favor of black students whose grades were lower. The case was expected to result in a landmark decision, as the judgment could have a bearing on all affirmative action programs wherein minority groups are given preferential treatment under quota systems. Bakke won his case, but the decision did not settle the question of the right of all affirmative action programs. There will probably be other such cases in the future that the Supreme Court will have to decide.

Cases in which the subject is not confined to one state will be tried in the federal courts. A classical case of this occurred several years ago. It lasted for a considerable period and required a great expenditure of time in preparing and presenting evidence. The dispute was over the rights to the waters of the Colorado River. The contesting parties were the various western states that comprise the Colorado River drainage system.

The federal courts have jurisdiction to hear civil cases between citizens of different states if the claim is more than $10,000. In most cases in federal courts, the case is tried before a judge or judges (Supreme Court); however, in certain types of cases there are provisions for a jury. Even patent infringement and copyright cases have been tried in federal courts before juries.

The individual states have their own judicial system in accordance with their state constitution and laws. In general the state courts are divided into categories of trial and appellate courts. The trial court is frequently called the court of general jurisdiction and is the first court to hear all types of cases from crimes to divorce, personal injury (liability), and contract disputes.

A judge will preside over the trial court and there may or may not be a jury. Where a person's freedom is at stake or in liability suits, a jury will likely be passing judgment. The trial court will generally be held at the county seat.

There is usually an appeal possible from a decision of a trial court. Appeals from a state court of general jurisdiction are made to an appellate court. In most states the appellate court is an intermediary between the trial court and the state Supreme Court. In states without the appellate courts the appeal will go directly to the Supreme Court of the state.

Most states have other courts that are subordinate to the trial court or court of general jurisdiction. These courts are established by state legislatures and handle minor cases and offenses. Municipal or city court is one type of subordinate court. Small claims court is another minor court where monetary redress of a minor amount is sought. In the small claims court the persons involved do not have lawyers and no jury is present. The judge makes the judgment.

There is one other special type of court that some states authorize. This is a court presided over by a Justice of the Peace. He is usually elected and frequently has no legal degree. He handles minor cases such as traffic violations, shop-lifting of a minor nature, etc. For many years the pay of the Justice of the Peace was a percentage of the fines collected. Under this system the defendant may be at a disadvantage. This system did produce very active campaigning at election time. Recognizing the inequity of such a system, most states have changed the pay of the Justice of the Peace to a fixed salary.

There are now special courts and judges that handle juvenile cases. Since a high percentage of juveniles brought to court (nontraffic) are first offenders, it is important to have judges who can supply some counseling and advice rather than just mete out justice.

State court judges are appointed by the governor or elected. Some states have a plan whereby the appointed judges must have their names periodically placed on the ballot for ratification. The vote is usually yes or no. If there are a majority of no votes the judge is then removed from office. Someone is then appointed in his place. In this system judges are not competing against each other for votes. The basic reasoning for this system of judge selection is that it is difficult for the electorate to make a quality judgment at the polls between two candidates for judgeship. The governor can make a selection upon the advice of the legal profession. However, the people have the right to remove the judge from the bench if his performance is not satisfactory to the people. This plan does have merit and should possibly be considered for a number of local elected positions which now require too much time and expense in periodic political campaigns.

E. JUDGMENTS

In criminal cases, the defendant is the person or organization that is accused of committing the specified crime. The plaintiff (the party seeking redress) is the state. The attorney for the state is the prosecuting attorney (sometimes titled District Attorney). The defendant is always entitled to an attorney who will represent and defend him at the bar. If the defendant does not have the resources to pay an attorney, the court will appoint one and he will be paid out of government funds. The defendant may have a choice of selection from two or three court-appointed "public defenders," but he cannot select any defense lawyer. "Criminal defense lawyer" is one category of specialization in the legal profession. There are many categories of speciality in law, as there are in medicine and engineering.

In civil suits the plaintiff and defendant are usually both private individuals or organizations, although a public officeholder or a government organization could be either plaintiff or defendant. This could very definitely be the case in a construction contract suit if a government unit was the owner of a facility being constructed.

In the early history of the country a government unit had sovereign immunity from suit by an individual or private organization. The United States Federal Government still retains this sovereignty except for certain types of claims. In 1887 the Tucker Act allowed the Federal Government to be sued for breach of contract. The Federal Tort Act, passed by Congress in 1946, permits lawsuits against the United States for certain types of legal wrongs. However, there is a special Court of Claims to hear cases against the government. If the Court of Claims decides that the government (or its agent) was at fault and the plaintiff had a legally justifiable case, it would recommend redress to the Congress. Congress would then have to authorize payment to the plaintiff. Sovereign immunity for state or municipal governments has led to many miscarriages of justice. It has also led to negligent action by government employees. In recent years many state legislatures have rescinded the law of sovereign immunity and allow all government units to be subject to lawsuits. The rationale behind this change is that it is more just that the loss to the injured party be shared by the taxpayers than only by the individual who was injured. There is also the opinion that if the government can be sued, government bodies will be more diligent in protecting the citizen from injury by negligent action. The rapid repair of roads and marking of dangerous sections of highways as well as more careful police action are in the best interest of the authorities if sovereign immunity is not granted for negligent action in such or similar hazardous situations.

Sovereign immunity was enacted with the concept that it prohibits judicial intrusion with the running of government. There has been a considerable lessening of this concern in the latter half of the twentieth century. The whole problem of civil rights and the many court actions this has produced has led judges from the U.S. Supreme Court down to lower courts to mandate certain requirements in the management of government functions. In recent years federal judges have mandated many administrative actions, such as the busing of school children to achieve racial balance, upgrading the facilities and standards in prisons, and also whether and where dams, highways, and powerplants may be built. When Congress makes laws that are general in nature, such as laws pertaining to racial equality, environment, industrial safety, etc., then the result is that the door is opened to a multitude of legal suits testing the interpretation of these general laws and subsequent government regulations. It then follows, as the night follows the day, that the judicial system of the government will encroach upon the executive branch. This has happened and will continue to occur as long as Congress continues to write legislation that is regulatory in nature.

There was a case where a judge ruled that taxes must be raised to upgrade the prison system. Since the judge was not an elected official there is strong evidence of taxation without representation.

When construction or engineering contracts are poorly written, unforeseen circumstances may occur such as illegal activities taking place, or one of the parties being unable to perform; therefore, court litigation may develop. Experience has shown that in almost all cases when contract difficulties occur a strong effort should be made to settle differences out of court. Compromises may be necessary on the part of one or more parties to a contract. Being proud and stubborn can be costly. In the case of a disputed contract where advice from counsel is asked, the contracting party should warn counsel that court action is a "last resort." Being confident that "one has a good case" may lead to great disappointment. The work required under the contract will most likely come to a halt as soon as legal action is started. Court cases may drag on for years, and the project can lie dormant during this time period. This may be very costly to the owner as well as the contractor. The contractor will have to pay for labor and materials for which he cannot collect payment from the owner until the litigation is ended. The contractor may have to secure the partly finished project from damage by weather, vandalism, or theft during the long period of time when the court action drags on.

The contracting parties should keep in mind that even though they and their legal counsel are confident they have a very good chance of winning a legal case, the court may decide otherwise. There are many considerations that may enter into court cases, and even though a lawyer may believe he has done his homework well, there is usually great uncertainty in litigation. Some people may feel that court cases may be compared to basketball games in that there is always a winner. However, this is not necessarily true. The winners may only be the lawyers. The party who wins the court case may still have a net loss when *all* costs are considered. For instance, a company or government agency that gets a reputation for quickly going to court when a contract is not being performed to their satisfaction will have the bids on future projects inflated by the bidders to cover such an event occurring in the future.

Contesting parties in a construction contract dispute should remember that most construction projects are highly technical, and litigation will likely require expert witnesses at a considerable fee. Many judges and almost all juries lack the knowledge and competence to fully understand the technical aspects of the case. Therefore, the end decision is in many cases questionable.

Another very important factor to consider is that, if a lawsuit is won, can the loser of the judgment (defendant) actually pay? Winning a court judgment may be one thing, but collecting the judgment may be more difficult. This may be the case more frequently in lawsuits against individuals than against organizations. In construction contracts, the owner may be protected from inability to collect by requiring the contractor to furnish bonds (*see* Chapter 10). The contractor can be protected from default of owner by placing a lien of ownership upon the portion of the contract that has been completed as well as any land on which it is sited.

F. ARBITRATION

In order to provide for an expedient method of handling disputes and thereby avoiding litigation, the contract may specify that certain matters be resolved by third parties. In some cases the contract may state that "in case of question over any phase of the construction the decision in writing of the engineer (or architect) will be binding on all parties." Such an agreed-upon arrangement would apply in the situation where the engineer or architect is an independent entity from the owner's organization. In cases where the contractor would feel the engineer or architect would be biased in favor of the owner (since he is in effect an agent of the owner), the contractor may want an independent engineer or architect or an arbitration board to settle the dispute. There are formal arbitration organizations in many parts of the country who perform arbitration services for a fee. Details of mediation and arbitration procedures are given in Chapter 11. These arbitration boards will have professionals within their organization who will be familiar with the technical details of the work. Sometimes decisions rendered in arbitration are not agreed to by one of the contesting parties and the case then ends up in the courtroom.

Arbitration is not always the best route to follow in a contract dispute. There are advantages and disadvantages to settling construction disputes by arbitration. Some of the advantages are:

1. It can be less costly than court litigation especially for minor disputes.
2. It can be a much quicker means of settling disputes; again, especially so for minor disputes.
3. The arbitration board may be much more qualified to understand the technical aspects of the dispute than a judge or jury.
4. A case in court may be determined on legal technical rules rather than on the merits of the case. This would not be the result in arbitration.

5. In arbitration cases, lawyers are more likely to play a less important role than in litigation. The contestants presenting evidence
   are more likely to be the owner's and contractor's personnel
   rather than the lawyers. Expert witnesses can question each other.
6. Arbitration cases can be kept private, thereby avoiding any adverse publicity to either party.

There are disadvantages in going to arbitration rather than to court.
These disadvantages will be more pronounced where the differences
involve major points of conflict rather than minor ones. Some of the
major disadvantages of arbitration are:

1. The arbitrator may not make a fair or quick decision.
2. There is no appeal except to go to court. The court will then
   decide if the arbitration proceedings were in keeping with the
   contract terms, and not if the final decision was correct or fair. In
   litigation, if a party is dissatisfied with the decision of the trial
   court, there is the right to appeal to a higher court.
3. It may be difficult to bring all parties, witnesses, and evidence
   before the arbitrator. The arbitrator has no legal power of subpoena unless he has been given this by state arbitration laws.
4. Relevant legal concepts which might have a bearing on the case
   may not be fully understood by an arbitrator. This may lead to
   later legal problems.
5. An arbitrator may have personal biases because of his own technical background. He may not conduct a totally fair hearing because of this bias. When major disputes are involved it is better to
   have a hearing before an arbitration board than one person.

The question foremost in an engineer or architect's mind is not which
is the best route, court litigation or arbitration, but "how can I avoid
having to be involved in either process?" Careful preparation of all
contract documents, good management, and careful inspection will be
very effective in eliminating contract disputes. The expenditure of funds
for legal review of the contract before signing may save a great deal of
money for legal fees after the work starts. If there are questions with
regard to any part of the contract document (especially the plans and
specifications), there should be clarification in writing before signature.
If there is a question in one's mind on whether any bit of information
should be included in the contract, the best rule is *put it in*. Engineers or
architects who obtain their work through submitting low bids will more
likely prepare hasty and incomplete contract documents. Detailed specifications may be lacking and the best use of materials and design proce-

dures may not always be followed. This in itself may lead to contract disputes between the owner and the design professional as well as between owner and contractor. Bargains are not always bargains even in the world of engineering, architecture and construction.

REFERENCE

1. Great Books, Vol. 2, *The Great Ideas*. Encyclopedia Britannica, pg. 964.

7. THE LAW OF CONTRACTS

A. DEFINITION

A contract is a promise between two or more parties to perform their responsibilities as set forth in the details of the contract. Contracts, as agreements between parties, are for the betterment of all the parties to the contract. Each party does something for the other party. The law of contracts is derived from common law in which redress can be obtained by one party of the contract for nonperformance on the part of the other party or parties. Since the majority of contracts are between two parties, the remainder of this chapter will be worded on the basis of two-party contracts.

A contract is a transaction involving two parties whereby each becomes obligated to the other with legal rights to demand the performance of what is being promised by each respective party. The promise made by both parties can be either verbal or written. A written contract, of course, is simpler to enforce and is less likely to lead to misunderstandings.

In order for a contract to be valid, it must meet criteria established by law. The following sections of this chapter will define the basic criteria of a valid contract which can be enforceable by a court of law.

Engineering contracts are usually *bilateral* in that both parties play the dual role of promisor and promisee. *Example:* Party A promises to build a bridge for party B in accordance with a given set of plans and specifications, while party B promises to pay party A a set amount of money when the structure is completed in accordance with all the details found in the plans and specifications.

Contracts can be expressed or implied. The majority of engineering contracts are *express contracts* in which all terms are declared by both parties either orally or in writing at the time the agreement is made. The above example of a bilateral contract is an *express contract.*

An *implied contract* is one which comes about as a matter of inference or deductions from facts and circumstances which can be proven in court to show a mutual intention to contract. Many engineers have been party to such contracts. *Example:* Party A asks party B to design the floor in his future garage so that he can have a storage room underneath. It is intended that party B will prepare the plans for this floor to carry the loads he sees necessary and that party A will pay party B for this work. If party B delivers the plans, then party A is obligated to pay the fee charged by party B. Party A is required to pay the billed fee unless he can prove in a court of law that the fee is unreasonable. It will be the responsibility of party B to determine the required loads and perform the design in such a way that party A will be satisfied.

Another example of an *implied contract* is when a person takes his car to a garage for repairs. The car owner and garage owner have entered into a contract in which the owner implies he will pay for having his car repaired.

Other types of contracts are *joint* and *several.* Such types of contracts involve several persons as one party to the contract and another person or company as the other party. In a *joint* contract, several parties join together in a united, combined action to execute a contract for the mutual benefit of all the parties. An example might be the building of a swimming pool to be used jointly by several families. In a joint contract, all the parties are individually responsible for payment to the builder of the pool. The pool cannot be divided into individual ownership. The contractor would want a *joint* contract so that if one of the parties could or would not pay their portion, the contractor could collect from the others.

A *several* contract is one in which several independent agreements have been expressed in a single legal instrument. A classic example is a contract entered into by several adjoining home owners to have a sidewalk built on their street. The contract would be several between the homeowners and the builder. Each homeowner would only be liable for the payment to the builder of that portion which is on or in front of his property. If one of the homeowners defaulted on his payment to the builder, the other owners would not be liable for the defaulted payment.

Some contracts are termed *joint and several.* In such all may be sued together or any one may be sued for full satisfaction to the injured party.

An *executed* contract is one in which the agreement has been performed by both parties and there are no remaining obligations on the part of all parties to the contract.

A *voidable* contract is a legal instrument but it can be voided if certain conditions do not develop. An example would be a contract wherein party A is to build a shop for party B. A clause is written into the contract that the contract is void if a building permit cannot be obtained. If the facility does not meet zoning laws and a permit cannot be obtained, party A cannot collect any damages from party B even though party A had expenses for getting ready to build. A form of voidable contract is one made with a person who is not legally liable to an agreement, such as a minor or a mentally incompetent person.

B. COMPETENT PARTIES

In order for a contract to be enforceable by law, both the promisor and the promisee must be legally capable of performance. In the first case there must be at least two parties to a contract. A person cannot contract with himself. This statement may appear rather obvious and a bit factitious. However, a person may act in several roles within an organization or he may be the contracting officer in several companies.

In general, a minor as a party to a contract may at his option void a contract with no obligation to the other party despite a loss by the other party. The definition of a minor is established by state law. Because of this voidability of a contract by a minor, most organizations will not enter into a contract with a minor. An exception to the repudiation of a contract by a minor is where the minor receives so-called "necessities." Necessities would have to be defined and ruled upon by a court of law in case of a voided contract. The law has required a minor to return goods or property obtained under a contract that is later voided. However, he is not obligated to pay for used, damaged, or lost property.

In order for a contract to be valid, both parties must be mutually competent. Each party must have sufficient mental capacity to understand the import of what he is doing. Mental capacity at the time of signing the contract is the determining factor in the validity judgment. Mental incapacity after signing is not a basis for voidability. As in the case of a minor, an insane person may be held liable for the reasonable cost of necessities supplied to him.

While a person who can prove insanity at the time of signing and then returns to sanity can void the contract, the other party does not have the same option of declaring the contract null and void. This one-way street of voiding a contract is the same as with a minor. A party cannot void a contract because the other party is a minor or mentally incompetent.

A contract made with a person who was intoxicated at the time of signing can be declared void by that person. He will have the burden of proof, however, that his intoxication rendered him incapable of fully understanding or comprehending the import of his actions.

In past history the contracting powers of married women were limited. However, this is generally no longer the case.

A corporation by its nature and in accordance with the terms of its incorporation can be a party to a contract. However, a corporation has limited powers of activity and contracts must be within the scope of its permitted activity. A contract outside the object for which the corporation was chartered is ultra vires and ordinarily considered unenforceable by the corporation. (Ultra vires means in excess of or beyond power authorized by law.) Persons entering into contracts with corporations should satisfy themselves that the corporation is acting within the powers of its charter as granted by the state Securities and Exchange Commission.

C. SUBJECT MATTER

Another basic need for a valid contract is for the subject matter of the contract to be definite and lawful.

Most contracts in which engineers/architects are involved have legal subject matter. All things not forbidden by law may be the subject of a contract. Agreements in restraint of trade or action that is in violation of antitrust laws are in general unenforceable. An agreement which cannot be performed without violating the law can be legally voided. An example would be where a contract is drawn to construct a building at a given location and upon applying for a building permit, the contractor finds that the facility would be against the zoning law. He can then void the contract since the action in performance of the contract would be contrary to the law.

It is quite obvious that there is a need for each contract to be clear and definite. The courts cannot enforce a contractual agreement which is lacking in certainty. All obligations must be completely described and the extent of the work in construction contracts should be clearly shown in words and drawings. Indefinite contracts are quite likely to end in conflict between the contracting parties, and may have to go to a court of law before settlement. It can be said that a contract that ends up in the courts is usually a poorly written contract.

D. CONSIDERATION

Consideration is what each party to a contract receives from the other party for completion of performance. The consideration in engineering/ construction contracts is usually money on the part of the contractor and a facility on the part of the client or owner. In the contract between client and design professional (engineer/architect) the client promises money as *consideration* for the design drawings and specifications of a facility desired by the client.

A promise to do what is obviously a legal or physical impossibility is not a legal *consideration* and will not support a binding promise in return.

The fairness of an agreed exchange is legally irrelevant. The courts will not determine if the *consideration* of both or any one party to a contract is adequate, fair, or equitable. If a contractor contracts to build a facility for a definite amount of money, he cannot later make a claim for extra money because he did not receive a sufficient amount to cover his costs and profit. The law will not consider if the bid price was sufficient or not; to do so would invalidate the whole system of bidding. However, a very important point the engineer/architect and owner should be aware of is that if the extent of the work is increased beyond what was specified in the original contract, then an extra amount of money is a possible legal claim by the contractor.

When legal transactions are made in the conveyance of property and one party does not require compensation (such as among family members) then the wording in the contract document commonly reads "for the sum of one dollar I . . . ." This makes the transaction legally binding even though one dollar is a mere pittance as a consideration in comparison to the value of the property given. This consideration of one dollar then protects the receiver of the property from later claims by the giver or heirs of the giver. An important point of consideration is that the fairness of an exchange is legally irrelevant.

If a promisor reserves the right to cancel a contract at any time, *without notice,* he has really not bound himself to anything and then the second party's promise can be voided for lack of consideration on the part of the first promisor. There is no mutual obligation in this case. However, if the contract reads that notice must be given for cancellation of a contract there is a binding contract until the notice is given. An example is as follows: Firm A agrees to supply builder B ready-mix concrete at $25 per cubic yard but reserves the right to cancel ten days after written notice. The contract is legal and binding upon firm A for the period of time up to the time of delivery of the written notice and for ten days after receipt by firm B of the written notice.

Another important factor in the subject of consideration is that any consideration must be definite in terms of its value. This point is important to the engineering profession. A builder's promise to build a "building" is not sufficient in terms of it being a definite consideration for the owner to promise the agreed-upon price. The building must be defined by plans and specifications. However, for the contract to be binding, the plans do not have to show every bolt or reinforcing bar in the building. Whether the description of the building was sufficient to make a legal consideration would have to be decided by the courts if the builder wished to void his contract. Without a clear description of details in the form of words and/or plans, the owner will be at the mercy of the builder on the quality and possibly the quantity of the details in the building.

Another rule of consideration is that doing what one is already legally bound to do is no consideration. A seller of property might contract with a buyer of a subdivision lot to install a sidewalk for a given price over and above the advertised price of the lot. However, it is later ascertained that the city requirements for approval of a subdivision is that the developer must provide paved streets and sidewalks. The purchaser of the lot can then void payment of the extra money for the sidewalk because he would receive no consideration since the developer is already bound by law to install sidewalks in the subdivision. The buyer of the lot then will receive no consideration for the money he would pay to the builder of the sidewalk.

A very famous contract case which has set legal precedents is that of *Foakes vs. Beer*[1]. The verdict rendered in this case was that "a lesser sum of money cannot be a legal consideration for a greater sum of money." In essence this says that the payment of a given debt by a lesser amount will not be binding even though the creditor is willing to accept the lesser amount as payment in full. The creditor or his heirs could still demand payment in full at a later date contrary to the former agreement. In more recent times this 1884 ruling has been repudiated by several court decisions. The voidance of the Foakes rule by the courts hinges upon the circumstance (which by legal considerations is technical) that the creditor did receive something as consideration over and above the lesser amount of money. One such technical consideration is time when the lesser sum is given before the due date of the debt. The lesser sum can be proven by economics to have greater value today than in the future. Subjective value of a "lesser sum today" has been accepted as legal consideration by the courts. Another approach to avoiding the Foakes rule is to give some material object in addition to the lesser sum. The courts do not, as previously mentioned, establish the value of a non-monetary consideration.

An illegal consideration which is counter to the law will be grounds for the voiding of a contract. Partial illegality of the *consideration* also renders the contract unenforceable.

There are certain types of contracts that appear to be without consideration but have been enforced by the courts. These types of contracts are called *Promissory Estoppel*. One example of this type is a gift promise. If A makes a promise of a gift to B, although B gave or made no consideration in return, the promise of A can be enforced, under certain conditions. If B has taken action on the basis of the promise so that if A defaulted on the promise B would suffer loss (time, money, etc.), then the courts have ruled in favor of B even though B neither gave nor promised a consideration. The basis for such a ruling is that injustice can be avoided only by enforcement of the promise to make a gift. Promises of charitable contributions are examples[2] of promissory estoppel.

A second case where the principle of promissory estoppel has been the basis for enforcement is the promise of a subcontractor to do work for a stated amount of money. If contractor A promises to give the work to subcontractor B if he wins the contract, then there is consideration on both sides and a valid contract. However, contractor A does not have to make the promise of the subcontract work to B. If A, using B's stated price, prepares his bid on the basis of the cost of the subcontract work quoted by B, then B is required by law to honor his price quotation. Subcontractor B cannot raise the price after he determines that A was the low bidder. Contractor A is not required to give the work to B and sometimes contractors will go out shopping for a lower price after receiving the contract. This is called "bid shopping" and is considered unethical in the construction industry. A contractor who practices bid shopping can soon find himself without subcontractor quotes on future jobs.

E. ORGANIZATION OF CONTRACTS

Every day thousands of contracts are made and substantial numbers completed. A very high percentage of all contracts made come to a successful conclusion. Only a small percentage ever find their way into litigation in a court of law. Almost every individual enters into several contracts of one kind or another during his lifetime. When one goes to a dentist and sits down in the dentist chair, one has entered into a verbal contract with the dentist. The engineer working in construction will be involved with contracts. Whether large or small, each contract should be carefully prepared. The chance of a carefully prepared contract ending up in court is very small.

Almost all construction contracts result in considerable sums of money as consideration. Any disagreement after the contract is signed can result in considerable loss to the parties of the contract. Disagreement can be avoided by contracts that are definite, clear, and in all respects above-board. Concealing any information from the other party in an attempt to gain an advantage is not only unethical but is very foolish. Concealing of information known by one party to the detriment of the second party is grounds for voiding a contract or for a judgment being made for extra compensation.

Construction contracts are usually prepared by the engineer with cooperation or review by a lawyer. The engineer has to detail the scope of the work and the provisions and terms governing the performance of the builder. The lawyer would review the agreement of the parties and the general terms by which the obligation of each party to the contract is defined.

It is obvious that construction contracts of any scope should be in written form. The written contract is composed of the following three parts:

1. general conditions;
2. drawings and written specifications;
3. agreement between owner and builder.

There are standard contract forms available to the writer of a construction contract. References 1 and 2 give such forms and also give reference to others. Standard forms should be modified to fit the specific conditions of the work to be done.

All special conditions pertaining to the work to be done should be carefully detailed. Conditions may be *express* or *implied.* Unless the implied condition is rather obvious, the condition should be expressed in the contract. The condition that the builder should conform to OSHA regulations of safety is implied in law and inclusion of such regulations or even mention of such would seem unnecessary. However, a good rule to follow is that if there is any doubt to the necessity of including any condition statement, clarification of subject matter, etc., then include it.

A final note with regard to written contracts is that there are five types of contracts that must be in writing in order to be enforceable:

1. contracts dealing with the sale of land;
2. contracts requiring more than one year for performance;
3. contracts for the sale of goods over five hundred dollars;
4. contracts to answer for the debt of another;
5. marriage contracts.

Legislative authority has deemed these five types of contracts too important to be left to the uncertain testimony of witnesses. Written contracts are a deterrent to fraud.

F. PERFORMANCE

Performance is defined as a party to a contract completing the requirements of the terms of the contract for which he is responsible. When performance is met by all parties then the contract is complete and is terminated. However, as will be discussed later, contracts can be terminated before complete performance without breach. In a contract, performance is met or a breach of the contract exists. Breach will be discussed in the next section.

In construction contracts performance by one party is almost always conditioned upon performance by the second party. For example, the builder must execute the construction before the owner is required to pay the agreed-upon money. Usually the conditions of larger construction contracts are that the builder will be paid in increments upon certification by the engineer or architect that a portion of the work is completed. If the builder does not perform then the owner is discharged from his obligation of performance—namely, making a payment of money. Since the builder's obligation should precede (in time) the owner's duty, it is called a *condition precedent.*

At times a difference of opinion arises as to whether a contract is complete and a party or parties have complied with their contracted performance. There is rarely any question as to whether the performance of payment of a sum of money has been met. However, the performance of a contractor on a construction contract is not so easy to ascertain. It is possible that a contractor working on a large project could continually touchup here, move a rock there, and so on. The courts and reasonable parties to construction contracts, will look upon *substantial performance* as a precedent to performance by the owner. Substantial performance is defined as the essential completion of the project, although there may be small and rather insignificant deviations from certain aspects of the specifications.

The courts will enforce the right of a contractor to receive his payment if he has substantially performed, yet they will not require the owner to pay for something he has not received. Details on procedures under the conditions of substantial performance are given in Chapter 9.

G. BREACH

Breach is defined as failure to perform the requirements of a contract. In most contracts involving engineers or architects, it is the exception that there is a complete breach of the contract by one or more parties. *Willful breach* is indeed rare, and when it does occur and is judged such by the courts, no recovery is permitted by the party committing willful breach even though there is substantial performance. Abandonment of the work would constitute willful breach if no conditions beyond the builder's control caused the abandonment.

A contractor who does not supply the exactly specified material may be acting willfully (and foolishly), but this action usually would not be termed willful breach. He would be in a position of substantial performance and arbitration, or as a last resort, the courts would determine the recovery the owner should receive for the lack of total performance on the part of the builder.

A *material breach* is defined as that nonperformance which results in the injured party (usually the owner in a construction contract) receiving something substantially less than or different from that which was intended by the contract.

Many construction contracts are by nature divisible contracts. As such, they can be and are divided into parts. When the builder or a supplier of materials delivers on an agreed-upon unit of the contract, the other party is responsible to perform by paying an agreed-upon amount of money. If after several units of work have been completed and the builder cannot continue to perform, then he is in breach of contract only for the remaining work. The owner can only recover damage due to nonperformance of the uncompleted portion of the work.

The contract for the sale of materials as a divisible contract has special conditions. For instance, a steel supplier contracts to deliver 800 tons of reinforcing bar each month for a series of months to a builder. For each unit of 800 tons of steel the owner is obligated to pay the agreed-upon price. However, if the supplier only delivers 500 tons one month, the owner is not obligated to accept the lesser amount. He can declare a breach on the part of the supplier and seek damages. If he goes ahead and uses any or all of the 500 tons he will be obligated to pay for the steel used, but can still declare a breach of contract and seek recovery for any damages due to failure to receive the full 800 tons of reinforcing bar.

A party to a contract may, in advance of the time when his performance is due, determine that the other party will not perform his part of the bargain. The first party may declare by notice that his further duty is ended. An example of this would be when a builder learns of the bankruptcy or insolvency of the owner. The builder could legally cease work, since his payment for any future (and possibly past) work done would be in jeopardy. An owner could also declare the contract void and no further obligation necessary if it is clear that the builder can no longer perform. Some additional reasons for nonperformance would be that the subject matter was destroyed or a change in the law rendered the subject illegal.

H. TERMINATION

Contracts are terminated because of the following reasons:

1. Completion of obligations by both parties—the optimum reason for termination.
2. Breach by one or more parties to the contract.
3. Impossibility of performance by one or both parties. The death or disabling illness of a person responsible for performance. The destruction of the subject matter would fall into this category of impossibility.
4. Mutual agreement by all parties to terminate.
5. Illegality of subject matter or persons party to a contract.
6. Fraud on the part of one or more parties.

7. Failure of consideration. If the value of a consideration is destroyed or almost destroyed by an uncontrolled event, then the performance by the party owning the destroyed consideration is impossible. An example would be when the transfer of ownership of a building was pledged as the consideration by one party and the building burned down before the transfer took place.

Further discussion of termination procedures and legal consequences of termination will be discussed in more detail in Chapters 8 and 9.

NOTES

1. L. R. 9 App. Cas. 605 (House of Lords 1884).
2. *Beathy's Estate vs. Western College,* Supreme Court of Illinois 1898.

REFERENCES

1. J. Sweet, *Legal Aspects of Architecture, Engineering and the Construction Process.* St. Paul, Minn.: West Publishing Co., 1970.
2. D. S. Barrie, and B. C. Paulson, Jr., *Professional Construction Management.* New York: McGraw-Hill Book Co., 1978.

8. ENGINEERING CONTRACTS

A. CONTRACTS BETWEEN ENGINEER AND CLIENT

The employment position of most engineers can be divided into categories. The employment categories are:

1. the engineer is self-employed;
2. the engineer works for other engineers;
3. the engineer works for an industrial or government organization.

The first two categories of employment are most likely connected with consulting engineering (*see* Chapter 5). Consulting firms sell engineering services. These services take the form of feasibility reports, cost estimates, design drawings and accompanying specifications, and construction contract preparation and evaluation. Construction management and inspection of construction are also possible activities of consulting engineering firms.

When engineering services are required, a client (owner-to-be of the new facility) will select an engineering or architectural firm to do the work. If this is a new activity for the client, he may be in doubt about selecting the proper engineer. He may inquire of other firms who have employed consulting engineers or he may obtain names from the "yellow pages" or listing of engineers in publications. Engineers do not advertise but technical magazines have sections where firms can provide names and addresses together with listed areas of specialties. The seeker for engineering services can publish a notice in technical periodicals that they are seeking an engineering firm, listing the type of work to be performed. This notice could, and usually would, call for the submission of an experience resume and possibly a proposal (*see* Chapter 10).

The procedure is to review these resumes and possibly select a few of the firms that appear most qualified. It may be that the organization needing engineering services has an engineer or small engineering staff within their own organization who would be best qualified to make an evaluation of design professionals. The firms appearing to be best qualified could then make a personal presentation and give further evidence of their ability to perform the required engineering work [1]. From this personal contact a consulting firm is selected. An additional basis for selection may be the work load of the engineering firm and thus their ability to prosecute the work in an expeditious manner.

After the firm is selected a meeting is held to arrive at the specific items of a contract. Such items will include the following subjects:

1. the scope and extent of engineering work;
2. the starting and completion dates;
3. construction inspection;
4. responsibility for allied engineering such as geotechnical, surveying and mapping, electrical, and mechanical engineering;
5. the fee.

The engineering firm will have to have a full understanding of the first four items before the fifth can be set. The owner may be new at such negotiations and need advice regarding a reasonable fee. This can be obtained from experienced firms, publications, or technical societies. The American Society of Civil Engineers has suggested fee schedules that can be obtained by writing to Society Headquarters. This fee schedule is just a guideline and is not a set fee that is binding upon the engineer or the client. Fees for engineering and/or architectural services will usually run between four to fifteen percent of the cost of construction of the facility. Good engineering design alone cannot be performed for less than about five to six percent of the construction cost. Very preliminary design or feasibility studies may be less, and complex projects that may require research or original studies could cost considerably more. Several procedures that are used to arrive at a justifiable fee are described below.

Fixed Percentage. This procedure of arriving at an agreeable fee is one of the oldest and most used. It is simple to apply since the engineering fee will be a fixed percentage of the final cost of the constructed facility. This procedure has lost a great deal of support recently because there is a negative incentive for the engineer to produce an economical design. There is a disadvantage to the engineer if the client makes changes in the design conditions or requirements after the design is in progress. This may result in redoing an appreciable amount of the engineering with no increase in fee percentage. This may prove difficult and result in a breakdown in relationship between engineer and client.

Fixed Fee. The fixed fee method is a common method of payment for engineering services when the amount of engineering work can be clearly and specifically defined. Some public officials (and no doubt others) believe this fixed fee should be obtained by submission of closed bids and then the work given to the lowest bidder. As explained in Chapter 3, this bid procedure for engineering services was at one time contrary to the Codes of Ethics of ASCE, NSPE, and AIA. However, this is no longer the case since the U.S. Department of Justice action. The degree of expertise and thoroughness of the engineering work is directly related to the amount of money to be spent on engineering. Projects engineered for the least cost can end up as either unsafe structures or overdesigned and thus cost much more. The extra construction cost of a project that is designed cheaply can be much more than the entire fee for engineering services. Mr. W. H. Wisely, then Executive Director of ASCE, stated the case very concisely when he wrote in a letter to the Governor of Florida,

> In the overall costs of an engineering project the increment going to professional services in planning, design and construction management is relatively small—rarely exceeding ten percent for basic services. Yet it is the quality of such professional services that not only determines how well the project fulfills its function, but also produces the project at the lowest possible construction, operating and maintenance cost. It is in the latter costs that true economy in the public interest is to be achieved. Any apparent savings resulting from the use of cheap engineering will represent economy of the most illusory and fake nature.

It is not possible to exercise judgment on engineering capability when price is the prime consideration in selection.

It is only those ignorant of the engineering process who can believe that an engineered project will turn out the same irrespective of the amount paid for engineering services. Engineering by inexperienced and novice engineers usually has a lower price than by experienced, top-

quality engineers. Quality engineering should result in a more functional, lower total-cost structure.

The fixed fee should be determined by the engineering firm through
a process of evaluating the total engineering required, applying unit
prices to the various engineering tasks, and then obtaining the product
of time and unit price. Experience is necessary to arrive at a correct cost
figure. An overestimate is not fair to the client and an underestimate can
lead to the ills of cheap engineering.

Cost Plus a Fixed Fee. This form of payment is preferable when the
amount of engineering work cannot be clearly defined. In many projects, especially those that are different, it is not possible at the beginning to clearly define the required amount of engineering man-hours.
An example of such a project was one in which the writer was involved.
The project requirements were to rehabilitate a manufacturing plant
that had been on a standby basis. The engineering work was to inspect a
multitude of buildings and the process equipment, write work orders to
the construction contractor telling him what work was necessary, and
also to inspect the finished work. In many instances equipment had to
be dismantled by the contractor before the engineer could pass judgment
on the extent of the necessary replacement or repair work. In this project
there was no way a fixed fee for the engineer could be obtained. Therefore, the engineering firm was paid a given amount of money per man-
hour plus a fixed fee that would cover overhead and profit. The fee per
man-hour would cover engineers and office staff salaries plus operating
expenses. In such a project it is very necessary for the engineering firm to
keep accurate account of time employed in the execution of the engineering work. Money has to be expended on this activity as well as on the
engineering itself.

In some projects the cost of the engineering services is determined by
the actual amount paid for salaries, fringe benefits, retirement allowances, and actual operating costs. In addition to these costs the engineering firm is then paid a percentage of these basic costs for overhead expenses and a fee for profit. When the client is the federal government, the
overhead expenses are established by the government through audit of
the engineer's records. This last procedure for paying engineering services has been the pattern used for many years for government research
grants. When the research organization is a university, then the government pays the actual labor cost, supplies, and operating costs plus the
established overhead.

In the above described procedure the client will then pay the "true" costs of the engineering. When the costs between hiring one firm or another are evaluated, the only point of difference would be the charge per man-hour for labor. The selective evaluation process of comparing, say, the cost between a firm that charges $50 per man-hour and one that charges $55 is not just comparing the two unit prices. The more costly firm may have more experienced and qualified engineers on their staff and not only will produce a better and less costly constructed facility but may also do the engineering for less expenditure in man-hours.

The client should first carefully evaluate the past experience and repu- tation of the engineering firm or, more importantly, the actual engineers who will be working on the project. The second consideration should be cost. A reputable, established engineering firm will perform the work for a fair and equitable price.

When the client is billed on the basis of costs per man-hour, the engineer must have a bookkeeping procedure that correctly keeps a record of the man-hours expended. All personnel must be carefully instructed in how to keep account of their time, and they should be diligent in prosecuting the engineering work. Managers should be expe- rienced and well qualified so that they can organize the work to accom- plish the most in the minimum amount of time.

Case 2 in Chapter 4 describes a situation where an engineering firm defrauded a government unit on a cost-plus-fixed-fee contract. In this case they charged the time engineers worked on another project to the government project. Such action is not only unethical but also criminal. A felony was charged and a principal of the firm convicted.

After an engineer has been selected by the client and a fee established, then a contract is drawn. In Chapter 7 the types and essential parts of a contract are given. The engineer–client contract can usually be quite simple in nature. The client may seek legal advice in drawing up the contract. The specific subjects that the contract should include have already been mentioned. One item of the fee schedule should include the procedure for the payment of the engineering fee. If it is a small project the fee is likely to be paid at the end of the project. The engineering firm will then have to finance the engineering costs through the project. If the project involves many man-hours of work over several months the client will usually pay the engineering fee in installments. This will definitely be the case of a cost-plus-fixed-fee contract. In this latter form of contract the client would usually be billed by the engineer on a monthly basis.

Some technical societies have contract forms that can be used by the client. Such forms are usually applicable to small jobs but can be expanded to fit larger, more complex projects.

The contract should carefully spell out what responsibility, if any, the engineer has in the construction phase of the contract. This can be important in any liability claims that may arise out of the construction work. It must be remembered that the engineer's contract is with the owner and he is acting as the owner's agent when dealing with the contractor.

In commercial building projects the engineer (whether he is in structural, mechanical, or electrical engineering) will most likely be contracted to the architect, who will have the design contract with the owner. The engineer will then be a subcontractor. It will again be important to clearly define the engineer's responsibility in the construction phase, especially his duties as an inspector of the work of the contractor. When negotiating the fee it should be clearly stated in the engineer's contract how much inspection is required, and the fee should reflect the extent of inspection.

A common occurrence in engineering projects is termed a *joint venture*. A joint venture is when two or more engineering firms join together in seeking a contract with an owner. If successful in obtaining a contract from an owner to perform engineering services, then the owner becomes the client and the joint venture becomes the second party to the contract. Joint ventures in engineering projects provide a wider range of talents and spread the financial risks and burdens. Where the project is large and complex a joint venture of two or more specialist firms may be more qualified to perform the engineering than any single firm. The joint venture should have an organization such that the client's relationship to the engineering work is as if dealing with a single firm. The joint venture should not be confused with the situation where the owner has separate individual contracts with several firms who are all working on the project individually. In this latter situation the owner will be responsible to coordinate the work and monitor the project to prevent overlapping of responsibilities or negligence of any phase of the work. The role of the client is much easier when a contract is with a joint venture than with multiple contracts to several engineering firms.

Joint ventures do have to be closely organized. They create conditions of coordination that are not present in single venture organizations. A joint venture agreement should be drawn up giving responsibilities of each organization. Division of profits and losses must be clearly defined.

When all possible questions are clearly answered in a written agreement then the organization can proceed with greater efficiency and smoothness and future legal problems will be avoided.

In recent years the engineer has been faced with problems that were rarely encountered two decades or more ago. These problems have been the result of great public concern over environmental issues. In many instances the engineer has been the primary target of such brickbats thrown by environmental groups. Direct confrontations between engineers and environmentalists have usually resulted in both groups looking down their own tunnel with no meeting of the twain. This is primarily due to the fact that neither group speaks the other's language. Most recently, however, there has been a better meeting of the minds, since engineers are now aware of societal concerns and the majority of the public realizes that no new construction or development can be absolutely 100 percent free from a possible disaster. Only the very extreme are blind to these facts.

The engineer is still placed in a position that takes skill in meeting all issues. He is hired by the client and his livelihood depends on the needs of the client. The true professional will not do everything his client asks of him. He may be faced with a commitment to the public and to his client [2]. His basic loyalty should be to the public. Where the wishes of the client clash with the public interest it is the responsibility of the engineer to educate the client to the needs of the public. This may require great skill beyond that of a technical specialist.

An example might be where an engineering firm is to design a storm drainage system for a city. The mayor is disturbed at the estimated cost of the drainage system and tells the engineer to decrease pipe sizes. The cost of the properly designed system will require an increase in taxes which may jeopardize the mayor's chances of being reelected. The engineer knows that a reduction in the size of the pipes will result in an inadequate design and may require very costly remodeling at a later date. Should the engineer consent to the wishes of the man who hired him or should he take his case to the public? The client–engineer team may tend to promote economy over conscience. This might be the case more often when the client is a private entity. The engineer is expected to serve two masters, his client and the public. Not only the Bible, but also history has shown the difficulty of such a position. In serving the client he may be required to consider almost solely the economic aspect. He can register his dissent to the style, esthetics, and scope of the project but in the end he will most likely have to bow to the wishes of the client. The bounds of the project may not be in the best interests of the public but this conclusion may be hard to determine. In arriving at such a conslusion there may be personal biases that have come to bear on the subject. The role of the engineer may not be easy when he is faced with the dilemma of service to two masters.

A difficult situation that is sometimes presented to consulting engineers and architects is the contingency contract. In many cases there is no written contract, only verbal agreements. The agreement for engineering work may consist of a promise of a written contract upon certain events coming to pass. Typical events that may consumate contingency agreements are successful bond elections, granting of loans, federal or state grants of money, or political elections. The owner may want to get an early start on a project, yet because of lack of final action on funding the project the client is reluctant to agree to a design contract. The design professional may need work and want to obtain a contract for the design work. This set of conditions can induce the design engineer to agree to begin design work knowing that payment for the work will only be forthcoming if the contingency event is positive. The design professional may be willing to take the risk knowing that there is a possibility of expending money for salaries and supplies for which there might be no repayment. He may believe the odds are in his favor.

The consulting design engineer may believe that the agreement to begin work early will place his firm in a favorable position with the client and may open the door to future contracts.

Besides the risk of expending funds for which there may be no reimbursement, the design engineer places himself in a compromising position. History has shown that design engineers have sometimes been badly damaged by contingency agreements. Clients have at times taken advantage of the engineer. With no signed agreement, a client with questionable ethical standards may try to drive a hard bargain. If there has been no firm agreement on the design fee at the start of work, a client really has the design professional in a weak bargaining position. In some cases a client has refused to honor an agreed-upon fee and offered a contract at a reduced fee. Since the agreed-upon fee was only verbal, the designer as well as the client knows it would be difficult to prove the conditions of the verbal agreement in court. On the other hand, the designer may lose all if he holds out for more money. Contingency agreements have led to bribery and payoffs. Dishonest public officials have taken advantage of the compromising position of the designer and have demanded payoffs for a signed contract. (*See* Case 3 in Chapter 4.)

Experienced and astute design firms usually have a general policy against becoming involved in contingency agreements.

In developing a design the engineer will develop a considerable volume of design calculations. In some types of projects, especially where the owner is the government, the owner may require a set of "as designed" design computations for review and his permanent records. In this case the computations may be more detailed, complete with explanatory notes and theoretical development. In any case, all the design computations should be assembled at the completion of the design and bound into a finished set. These computations should be preserved by both the design professional and owner for any possible future reference. They could be used as evidence in any legal action brought against the design professional. Because of this possibility, and as protection in case of a suit, the design computations should be thorough, complete, and in accordance with the latest knowledge of the profession and the latest design codes and specifications.

B. INVOLVEMENT OF THE ENGINEER IN CONTRACTS BETWEEN OWNER AND CONTRACTOR

In order to build any facility, it is necessary to have a signed, written contract between the owner and the builder (contractor). Although the engineer may be very involved in preparing the contract documents and also in evaluating the bids submitted, he is not a party to the contract. The engineer is an agent of the owner and should see that all terms of the contract document are consistent with the plans and specifications.

The amount of involvement of the engineer in the negotiations between owner and contractor will depend on the status of the owner. When the owner is the federal government, the engineer's role in this phase of the project may be very minor. However, when the owner is a private organization, the engineer may have the major role in drawing up and overseeing the contract between the owner and contractor. The engineer may be the project manager and have the role as construction manager. He will then act in the capacity as agent to the owner and will oversee the work of the design engineer as well as the builder.

When an engineer has a contract with a client to provide engineering services, the engineer/architect becomes the agent of the owner. The role of an agent has legal implications and restraints. Both the client and engineer should be knowledgeable of the role of an agent. The engineer as agent represents the client (who is called the principal) in dealings with third parties. The agent must have the authority from the principal to act in this role and the limits of the agent's authority should be contained in a written document. An agent has a differing role from that of an employee. Usually the agent has more freedom of action than an employee and also more responsibility. There may be cases when an employee would act as an agent of the employer.

The agent acts not for himself, but for the principal and under the general control of the principal. Action that the agent takes within his authority is binding upon the principal. When a third party deals with an agent, he should be sure the agent has the authority and he should also know the extent of the agent's authority. The principal is not bound by an agent acting outside his authority. The agent can place himself in legal jeopardy if he acts outside his authority. He would be charged with fraud by either the owner or a third party.

The agency concept is very important in the business world. Officers of a corporation are agents of the corporation. A contract of partnership is a contract of agency. In a partnership each of the partners serves the double role as an agent and a principal. The acts of any one partner are binding upon the other partners.

The authority to act as an agent has all the requisites of a contract. There must be mutual agreement between both parties. The activity and responsibility of the agent must be legal. *Power of attorney* is a form of authority. A *proxy* is frequently used in corporate business activities and is another form of written authority.

The principal may create an agency by implied action even though he has given no written authority to the implied agent. Such agency is given the legal terminology *agency by estoppel*. A person may represent another in a business action even though he does not have the authority to do so. The action may be favorable to the person being represented. Such acceptance is *agency by ratification*. This may give the appearance to the third party and even to others that there is a principal-agent relationship. The principal has in this case knowingly given the agent "apparent" authority. The agent might then operate in other transactions as an agent even though he has no real authority. The principal may become (as a result of court action) fully liable to third parties for the agent's actions if it can be proven that the actions of the principal were such as to imply agency.

An agent has certain duties and obligations toward the principal. He is expected to utilize reasonable care and loyalty. His actions must be ethical. He should keep the principal informed of all actions. Unauthorized acts of an agent may be ratified by a principal. If acts are outside the agent's authority and are not ratified by the principal, the agent is subject to a suit of tort liability or fraud.

In nearly all construction contracts there will be several construction firms working on the project. This will probably not be a joint venture but the project will have a *prime contractor*. The others will be subcontractors to the prime contractor. The subcontractors will have contracts with the prime contractor, not the owner. In some projects the terms of the contract between the owner and the prime contractor require approval of the subcontractors by the owner. The purpose of this approval clause in the contract is to assure that the subcontractors are qualified and are capable of handling the work in a satisfactory manner. Any subcontractor that has performed in an unsatisfactory manner on previous jobs can be screened from the project. The screening of subcontractors generally presents no legal entanglements where the owner is a private organization. However, when the project is to be paid from public funds, the nonacceptance of a contractor or subcontractor must have solid justification. It cannot be done at the whim of an individual or on the basis of personal dislike. Incompetence, lack of experience, or poor financial condition are grounds if they can be proven. There must be good evidence that a contractor or subcontractor is not competent to complete the work in a satisfactory manner before they can be excluded from a public project. A project in which private funds are used has more freedom of action. Nevertheless, the law allows no discrimination on the basis of race, creed, or sex in any business action.

There are also joint ventures in construction contracts. Multimillion dollar contracts may be more than a single contractor wishes to handle or finance. A new construction firm may be formed in which the backers and holders of stock in the new company will be several contractors. In this way more expertise can be made available and any risks or future profits will be distributed among the consortium. The building of Hoover Dam in the 1930s was one of the early uses of this joint venture procedure. The contract for the building of this large dam was between the U.S. Government (Bureau of Reclamation) and Six Companies (the name of the joint venture builder). The joint venture, as the name implies, was formed by six construction companies. Six Companies was an entity only for the construction of the Hoover Dam. After that project, the six firms went their separate ways in the construction business.

The selection of a builder can be done in several ways. When public funds are used, the standard procedure is to award construction contracts on the basis of closed bids. Details of this procedure will follow later in this chapter. Most government units have laws requiring that selection of the contractors be on the basis of closed bids.

A private owner has the freedom of selecting a builder in any manner he desires. A closed-bid system is commonly used for large projects. However, a contractor may be selected on the basis of reputation of its past work, and a construction price is then negotiated. The price for the work may be on the basis of cost plus a fixed fee or on a lump sum payment. The lump sum payment is likely to be used on small projects when the owner has information from his engineer or architect on an estimated cost. The owner may negotiate with the selected contractor around this estimate. If there is no meeting of the minds on price, the owner may go to a second prospective builder. The details of operation of the various payment procedures are given in Chapter 9.

The owner, engineer (or architect), and the contractor-builder form a triumvirate in the construction process. The engineer is considered an agent of the owner and the authority of the engineer in the construction phase should be carefully defined in the contract between the engineer and owner. The channels of communication between the three organizations can take several forms. These channels are shown diagrammatically in Figure 8.1. In the method shown in Figure 8.1(a) direct communication is found between any of the three entities involved in the project. In this method of management the contractor will have direct access to either the owner or engineer. His contacts with the owner would be on all phases involving financial matters, progress, and all

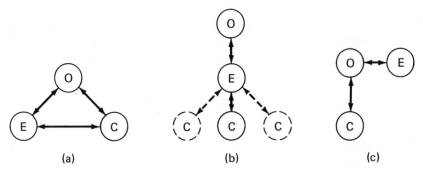

FIGURE 8.1. Various methods of owner–engineer–contractor relationship.

other business matters. The contractor would make direct contact with the engineer over technical matters, inspection, and progress reports. In this system, care needs to be taken to see that all three parties are always informed on any decisions that affect the work.

The management method shown in Figure 8.1(b) has the contractor communicating with the engineer on all matters and never directly with the owner, even though his contract is with the owner and not the engineer. This would be the preferred method where the project is large and involves several different prime contractors (other contractors are shown by dotted circles). In this situation the engineer would serve as construction manager.

The management method shown in Figure 8.1(c) has no direct contact between the engineer and contractor. In this situation the engineer usually serves as the designer of the project and when the construction starts, has the role only of answering technical problems and serving as advisor to the owner. This would be the arrangement in the situation where the owner has in-house engineering staff who would monitor the construction, and the inspectors would be employed directly by the owner. This arrangement is common on projects in which the U.S. Government is owner and the project is under jurisdiction of the Bureau of Reclamation or Corps of Engineers. State highway departments are most likely to operate under method (c) when they contract design work to private consultants.

Whatever the role the engineer is to play in the construction project, his duties should be clearly defined in his contract with the owner as well as the contract between the owner and contractor. This can eliminate problems after the work starts and could also prevent legal confrontations. It may also prevent later claims on the owner for extra compensation when either the engineer or contractor feel they performed services which were not included in the contract. A clear understanding of responsibilities will result in many benefits to the project and will help considerably in smooth and expeditious progress to completion.

C. EMPLOYEE–EMPLOYER CONTRACTS

Most employees upon entering employment do not consider the act of being hired to be a contract. It is very unlikely that the employer will think so either unless he employs many people. Yet the definition of a contract (Chapter 7) says the agreement of employment is a promise that can be enforceable by law. There is a promise made by both sides, the employer and the employee. Even though there is no written contract in most cases, if the parties are legally competent and the subject matter is legal, then an employment agreement is a contract. An exchange of letters can be considered a written contract.

The primary difficulty in enforcing such employment contracts is that there is no time limit to the contract and as such it can legally be terminated by either side at any time. There may be cases where the engineer may sign a formal contract document for employment. This will be likely if the place of employment is outside the United States or if he is assuming a top management position. Without a signed document setting forth the period of employment or verbal evidence of definite time period, the length of an employment agreement would usually be considered by the courts to be a pay period.

If a person accepted employment and went to considerable expense, such as moving to a new location, and then was terminated shortly thereafter, the employee would have grounds for a claim against the employer. A fair trial period would likely be considered necessary before termination. Termination of government employees has been wrought with difficulty. Civil service regulations usually require that termination can only be the result of dishonest behavior, incompetence, and the elimination of the position. In the latter case the employee has top priority on vacancies occurring elsewhere for which he is qualified. Declaring a person incompetent is a disagreeable action and can lead to a special hearing before a review board. Most supervisors will retain those with marginal performance rather than go through a hearing. Legislators and administrators in government are seeking means whereby those who are incompetent can be terminated more quickly and with less waste of time in procedures, yet still protect the individual against the unethical action of supervisors.

There is not much discussion and rarely is any action brought against unethical practices of employee or employer. There are statements in Codes of Ethics that pertain to the employer-employee relationship. In the ASCE Code of Ethics, Fundamental Canon No. 4 states, "Engineers shall act in professional matters for each employer or client as faithful agents or trustees." The IEEE Code of Ethics, in item 1 of Article 11, states, "Engineers shall, in their work, treat fairly all colleagues and co-workers, regardless of race, religion, sex, age, or national origin."

Item 6 of the same article states, "Engineers shall, in their work, assist colleagues and co-workers in their professional development."

The Guidelines to Professional Employment for Engineers and Scientists (*see* Chapter 3) cover quite completely the relationship between employer and employee and ethical conduct of both. Adherence to these Guidelines by both employer and employee will do much to enhance the profession of engineering, not only in the environment of the engineering office but also at the construction site. The efficiency of the engineer's work will be at a much higher level when the work is performed in a spirit of cooperation. A well run engineering office will strive to have a high morale among its employees. An employee who is disgruntled has an ethical obligation to the employer to make known the reasons of his discouragement, and then if conditions are not to his satisfaction he should change his place of employment. The new engineer should not look for perfection in his employment, but should be realistic in his desires. A positive mental attitude can do much to promote good relations between employee and employer. It will also result in better engineering. Creativity and excellence never came out of a mind cluttered with bad feelings and negative attitudes.

REFERENCES

1. G. L. Devlin, "A Statement on Professional Services—Competitive Bidding," *Engineering Issues*, ASCE, January 1975, pp. 53–59.
2. T. L. Turnick, "Public Versus Client Interests—An Ethical Dilemma for the Engineer," *Engineering Issues*, ASCE, January 1975, pp. 61–65.

9. CONSTRUCTION CONTRACTS

A. INTRODUCTION

Construction contracts are those agreements between an owner and a builder (termed the contractor) in which the owner agrees to pay a certain sum of money for a constructed facility. The constructed facility may be anything from a house to a multimillion dollar highway, dam, or industrial complex. The contract may vary from a few typed pages and one or two sheets of drawings to several books of printed matter and over several hundred sheets of drawings. Regardless of the size of the project, the basic contract concept is the same and in principle identical. All construction contracts involve promises on the part of both parties. Consideration is offered by both parties—the owner offers money; the contractor, a constructed facility.

The construction industry is the largest industry in the United States in dollar volume. Each year there are thousands of construction contracts signed and thousands of others successfully completed. There are a small percentage each year that develop disagreements and end in court. Such defaulted contracts are costly to one or all parties to the contract. The preparation of construction contracts should be carefully organized. Avoidance of legal complications should be one of the major objectives of those preparing the contract. The engineer or architect is responsible for the writing of the contract, preparation in consultation with the owner, and any legal help that may be deemed necessary. The engineer or architect must be sure that all the information necessary to build the facility in accordance with the wishes of the owner is contained in the plans and specifications. The design engineer has a moral responsibility to both the owner/client and to the contractor. He should not have the attitude that since he is under contract to the owners that he should use

125

all means to get as much for the owners' money as possible even at dire consequences to the contractor. This attitude, if fully exploited, will very likely lead to trouble on the job and may cost the owner legal fees as well as damages in settling disputes. Another result will be higher bid prices on future projects in which the engineer with the biased viewpoint is involved.

A major responsibility of the engineer or architect is to precisely determine the wants of the client. Sufficient time should be spent with the owner or his representatives to have a clear understanding of what the final project should be like. It may be that the owner only has a basic performance specification. In some types of projects this may be a very simple requirement. For instance, a state hires a consultant to design a highway bridge. The specifications to a consultant may be only that it is to span a certain river on a specific highway number. The roadway requirements are specified and so is the class of weight loading. The design consultant is then required to determine the best type of bridge. Alternatives most likely will be presented which will show feasible types of bridges with costs, maintenance evaluations, and esthetic considerations. This type of project would require a minimum of conferences between owner and engineer.

A more complex type of design project that would require much interchange between owner and engineer would be a large manufacturing facility. Many alternate designs would need careful study throughout the design development.

During the design development it may appear prudent to consult with the contractors to obtain their viewpoints on major features of the design and possible alternate features. This procedure would be more likely used on an unconventional or "first of its kind" project. Care has to be taken in consulting with future contractors when public funds will be used to build the project. Laws may rule out a contractor bidding on a project if he has advanced knowledge of detail information before the drawings and specifications are available for bid preparation.

A prebidding conference may be part of the bidding process. This conference would involve the owner, the engineers and/or architects, the interested builders (future contractors) and material suppliers. This conference would be called some time after the interested contractors have had time to study all the contract documents, including plans and specifications. In this conference the engineer/architect will explain the general features of the project and any special or unconventional aspect. The contractors can ask questions regarding the unclear items. The answers are then given to all so that there can be no claim of information being restricted to a limited number of possible contractors. Record tapes should be made of all the proceedings in such a conference for future reference in or out of court. A written addendum to the *invitation to bid* should be sent to all prospective bidders. This addendum should contain answers to questions submitted at the pre-bid conference.

In the present world of heavy construction in the United States a general contractor (the prime contractor) is primarily a construction "broker." His direct employees on a project may be only a handful of supervisory personnel. All the work may be done by many specialized subcontractors. On the construction of a large building there may be many subcontractors, from the excavation subcontractor to the floor-covering subcontractor. All possible subcontractors and material suppliers should be present at the prebidding conference.

The contract for the building of the facility will be between the owner and the general contractor. Direct contracts between owner and subcontractors are not likely unless the owner is acting as the general contractor and then deals directly with the "subs." The owner must then have experienced personnel who handle the construction supervision.

Although the corporate separation of engineering and construction is the common practice in the United States, some firms provide both services. These firms are called turnkey operations—meaning the owner deals with only one firm from the beginning of planning until the facility is entirely complete, and the design–construction firm turns the keys over to the owner. (Chapter 5 discusses the pros and cons of the turnkey type of operation.)

It is possible that an owner or engineer may divide the entire project into several parts. Separate contracts may then be drawn with prime contractors for each part of the work. If this mode of operation is followed it is best to be sure that the work on the separate parts can be clearly defined. There are dangers of friction and claims of interference when more than one contractor is operating at one site simultaneously. It takes careful management of the entire project by the owner or his technical representative. If one phase of the work can be finished before the next begins then multiple contracts are feasible, especially if total cost can be reduced thereby. For example, building construction would be the subject of one contract, the installation of equipment the subject of a second contract, and the roads, streets, parking, and landscaping the subject of a third contract.

After the design is completed by the engineer/architect the project is prepared to "go to bid." All the required documents are prepared. The work to be contracted is advertised in several ways. If the owner is a private firm, notice (*Invitation to Bid*) of the impending work may be sent to a select few highly experienced contractors. If the work is to be funded from tax or government bond monies, the law will likely require a public announcement of some form. Construction magazines will usually have a section for bid advertising. There are other periodic publications that list jobs to be let. A typical advertisement, from the December 14, 1978 issue of *Engineering News Record* is as follows:

> The Port Authority of New York
> and New Jersey
>
> Sealed proposals for the following contract will be received by the Chief Engineer, Room 605, One World Trade Center, New York, NY 10048, until 2:30 p.m. on the date indicated and will then be opened and read in room 60E. Contract documents may be seen at room 58E and will be furnished upon request. Questions by prospective bidders concerning any one of the contracts should be directed only to the person whose name and phone number is listed for the contract in question. No deposit is required.
>
> Contract LT-214-Lincoln Tunnel-Rock Bolting Along Helix: Bid due Thursday, January 4, 1979—Direct questions to Mr. M. Loriz (212-466-7196).
>
> The Port Authority of New York
> and New Jersey

The above sample of an official proposal is quite brief; many are in more detail. Official proposals covering work to be paid by the Federal Government in part or in whole today contain the following statement: "The Contractor will be required to comply with the federal equal employment opportunity regulations."

It is usual to require a money deposit in order to obtain a set of plans and specifications. This deposit will usually vary from $25 to $100. If the person (or firm) makes a bona fide bid on the work and is not selected as the contractor, then he can usually receive his deposit back upon returning the set of plans and specifications.

In order for a contractor to bid on a project (public or private) he should be qualified to do the work. In some projects there is a list of qualified bidders. In order for his bid to be considered, the contractor must be on the qualified bidders list. There are specific requirements for approval to be listed as a qualified bidder. The primary requirements are sound financial position, experience in doing the type of construction required, successful completion of similar projects of a magnitude nearly equal to the proposed project. It does not have to be a project of the same or larger magnitude since this would close the door to new contracting firms moving up in the magnitude of jobs they are permitted to bid on. Most owners have a prequalification procedure for prospective contractors.

B. THE BIDDING PROCESS

In most construction work there is a bidding process that takes place to determine which organization the owner will enter into a contract with for the work to be performed. Only in turnkey operations or where the private owner desires a particular construction contractor will there be no competitive bidding. In these latter two cases the contract will likely be on a cost-plus-fixed-fee payment plan. This payment procedure will be discussed in more detail later in this chapter.

A qualified prospective bidder, upon seeing an official proposal notice, will, if the company desires the contract, proceed to obtain a set of plans and specifications and all other information for bidders. It is very important that the owner make sure that all bidders receive all of the information. A written receipt should be obtained from each bidder verifying that each bidder has received all the prepared information. If additional information such as modifications or corrections to the plans and specifications are to be given to the prospective bidders by a written addendum, it is necessary to make sure they each have received a copy by returning a signed receipt for such. If a bidder does not receive a piece of information the other bidders receive and he is not the low bidder he may legally block the granting of the contract to the lowest bidder, especially if he can prove that if he had received the information it would have altered his bid price. The process of bidding would then have to be repeated, thus causing a delay and maybe higher bids on the second bidding.

As noted in the sample advertisement there is a deadline date for submission of bids. There are rarely any exceptions to this, especially for a public owner.

The contractor, after receiving all the preparatory documents, starts working to make a "takeoff" of materials and estimated costs. He will also have to estimate the cost of labor. This cost estimate will take time and staff in preparation. It will be necessary to double check all work to see that no mistakes have been made. Mistakes can be very costly. There are companies that specialize in preparing quantity cost estimates and also the man-hour requirements for all trades.

Almost all construction projects will involve a prime contractor making the total bid to the owner and then subcontractors making price bids to the prime contractor for their portion of the total contract. It is then necessary for the prime contractor to contact subcontractors and obtain their price proposals. The prime contractor will want to obtain several bids for each type of subcontract work. A building project may require subcontractors in excavation, concrete, reinforcing or structural steel erection, masonry, electrical work, mechanical work (plumbing, heating and air conditioning), glazing, doors, ceiling tile, floor covering, painting, landscaping, and possibly other trades. Bids from subcontractors are usually not received as formally as the total project bids, but such bids can and do form legal contracts and such subcontract price proposals should clearly define the scope of work and the price as well as time for the beginning and ending of the work. More details and ramifications on subcontracting are given in the next chapter.

After the prime contractor has received all his price quotations, he then prepares the official price table. This may be quite simple and short, or long and more complex, depending upon the type of project. There will usually be an entry column giving the work item, the next column giving the unit of measure, such as cubic yards, linear foot, ton, or gallons. Where it cannot be easily measured, the unit is lump sum (ls). The third column is the quantity for each item listed. It is very necessary to be sure the quantities from the work sheets are correctly transferred to the bid table. The fourth column is the unit price for each item. The unit price is for supplying all material and labor. It is commonly spoken of as the "in-place" price. The fifth and last column is the total price for each item. This is obtained for each item by multiplying the quantity by the unit price.

All items in the unit price column are totaled. To this total sum is added the contract extras for overhead, contingencies, and profit. This is then totaled to obtain the bid price.

Table 9.1 shows a typical price table, showing detail and method. This price table is combined with any other required processed forms, together with certificates of required bonds, and submitted to the owner or his representative at the designated location before the prescribed deadline date and hour as given in the proposal invitation.

The unit prices are used as the base price in case there is any change in the contract that would modify the quantity of work. In a lump sum

TABLE 9.1   Bid Price for Construction of Highway Bridge
Including All Labor and Materials

Item	Unit	Quantity	Unit Price	Total Price
Mobilization	ls	1	430,000	430,000
Superstructure				
Concrete, Class A4	cy	4744	130	616,720
Prestress Concrete bms	ea	203	6200	1,258,600
Epoxy	lb	181,420	0.42	76,196
Rein. steel	lf	762,190	0.23	175,304
Steel plate girder	ls	1	358,000	358,000
Bearing plates	ls	1	76,000	76,000
Concrete parapet	lf	3453	25	86,325
Median barrier	lf	3404	28	95,312
Drainage system	ls	1	30,000	30,000
Electrical work	ls	1	110,000	110,000
Elastomeric expansion	lf	453	155	70,215
Lumber—treated	mbf	55	2000	110,000
Substructure:				
Excavation	cy	2369	13	30,797
Timber piles	lf	30,130	7	210,910
Prestress conc. piles	lf	84,887	20	1,697,740
Load test piles	ea	4	8000	32,000
Concrete, Class A3	cy	1306	130	169,780
Concrete, Class A4	cy	11,240	100	1,124,000
Reinforced steel	lb	2,271,800	0.23	522,514
Riprap	sy	687	25	17,175
Concrete slope protection	sy	327	20	6540
Total Labor and Material Price				$7,304,128

contract this change in quantity is unlikely, since the project is a fixed entity. In some lump sum contracts the engineer/architect may list the items, unit, and quantity and then the contractor would just insert his unit price and extend the table to total price. In other instances the table is left blank and the contractor identifies each item and quantity. In the case where the items are identified and quantity listed by the owner or his representative, the contractor should check the work and calculate his own quantities. Any difference should be brought to the attention of the owner in time for corrections before bid due date.

When quantities are not listed, a contractor may approach the design engineer for a check on his calculated quantities. To prevent later charges of erroneous information (in case of an error by the engineer), this quantity information should not be given. Any information given to one prospective bidder must be given to all bidders by written addendum.

Unit price contracts are used when the precise quantity of work cannot be established before work begins. The items of work are listed and estimated quantities are also given so that total cost can be calculated. The quantities are clearly identified as estimated. The bidder knows that he will be paid on the basis of actual measured quantities multiplied by his listed unit prices. The quantities will be measured by the owner or his representative as the work progresses. The contractor will also verify quantities.

In unit price contracts the bidders will attempt to determine what the final quantities may be. If they can be sure the estimated quantity is likely to be low they will place a higher unit price on this item than they would otherwise. Item quantities that may be estimated high will most likely result in unit prices lower than normal. This manipulation of unit prices based on quantities is referred to as unbalanced bidding. It has, in rare cases, been cause for disqualifying a bid. The unit prices would have to be quite different from the engineer's estimate for the owner to have justification for labeling a bid unbalanced and disqualifying it. This is especially true when the owner is a governmental unit. Estimated quantities should be as precise as possible.

If the owner is a private entity then the bids can be opened at the discretion of the owner. If the owner is a public organization then the bids are opened in public at a designated time and place. A private owner may follow the same procedure. At the time of the opening of each sealed envelope the price is read aloud, shown to a witness and recorded. After all bids have been read then the low bidder is announced. It will then be announced that this bid will be examined in detail and if everything is in proper order and legally correct then contact will be made in the near future for the formal signing of a construction contract. The contract documents given to the prospective bidders along with the plans and specifications will usually contain a copy of the contract so that the potential bidder will know the complete wording of the contract before he submits his bid. In this way there are no unseen or unknown contract provisions introduced at the time of signing.

The law covering the jurisdiction of the public owner will most likely require that a contract be entered into with the lowest bidder if his bid has been in order. The official proposal may say "any or all bids may be rejected." However, a rejection of a bid on a public project must be justified, it cannot be capricious. State and federal laws or regulations require all bids be rejected if the lowest bid is a given percentage above the engineer/architect's cost estimate. A private owner can legally be selective but it is not ethical if he does so without good cause. He may find his reputation suffers if he does not accept the lowest bid. All bidders will have bad feelings (with the exception of the contractor who receives the contract) if the lowest bid is not accepted. There can be a considerable expenditure of time and money in preparing a bid and a contractor does not want to stake this cost upon capricious behavior on the part of the owner.

There are reasons that a bid can be rejected at the time of bid opening, or shortly thereafter and before the signing of a contract. A bid on construction must be "responsive to the invitation for bids." A bidder is not responsive when he does not conform to all the rules and regulations outlined in the *instruction to bidders*. Some of the ways a bidder may be nonresponsive are listed below.

a. He fails in filling out his price quotation properly by not doing his arithmetic correctly. Sometimes minor corrections are allowed such as correcting a multiplication in the product of unit price and quantity. He may not, however, change either the unit price or quantity. Further discussion on bidding errors is given in a later section.

b. He submits a bid after the owner's deadline. Generally such late bids are not accepted.

c. He places qualifications on the bid. Inserting a qualification into the bid that varies from the bid invitation will place that bid on a different footing from the others. Therefore, it cannot be allowed. Qualifications which have been inserted in bids, in the hope that they will be accepted, are:
1. change in time of performance;
2. "prices quoted are subject to no change by supplier";
3. bid as submitted is valid only if the contractor is not low bidder on another job that is to open within the same week;
4. a change in material specification.

d. The bidder fails to include a bid bond. A bid bond is a guaranteed sum of money that is required to be submitted by the bidder when he submits his bid. In case the bidder is the low bidder and then he fails to sign the contract the bond is forfeited to the owner. In past cases of bid openings a bid has been thrown out because the bidder failed to enclose a bid bond. A 1978 court case (*Lorenz vs. Plaquemines Parish Commission Council*; Court of Appeals of Louisiana, fourth circuit, 11/8/78) ruled that even though a road builder failed to submit a bid bond along with his original bid documents the bid could not be rejected. This ruling rejected a suit by the next highest bidder to have the award set aside because the low bidder failed to submit the bid bond. The low bidder had inadvertently failed to include his bond and did furnish the bond shortly after bid opening and before the signing of a contract. More information on contract bonds is given in Chapter 10.

e. Failure to list the subcontractors selected by the prime contractor. This requirement may or may not be in the instruction to bidders. An owner may have an approved list of subcontractors and all subcontractors to be used on the project must be on this list.

C. THE COMPONENTS OF A CONSTRUCTION CONTRACT

A construction contract will consist of several documents. These documents will consist of a number of parts. The organization of such is the perogative of the owner and engineer/architect. Through the years a general organizational pattern has developed in the contract documents. These documents generally consist of:

1. Agreement Form (between owner and contractor);
2. General Conditions;
3. Drawings;
4. Specifications;
5. Addenda.

In general terms, the agreement form defines the work to be performed by the contractor. It contains the date of the contract and the amount of money the owner is to pay the contractor. It defines the contract documents, names the engineer or architect, the time of starting and finishing the project, the manner of payment, and any special agreement pertaining to the project.

The General Conditions consist of all information necessary for the execution of the work that is not given in the drawings or specifications. The contents of the General Conditions typically consist of the following sections.

a. *Contract documents.* This section details what the documents consist of, the number of copies furnished the contractor and many other details.
b. *Engineer or architect.* This section defines the powers and duties of the E/A in relation to the construction.

c. *Owner.* This section defines the owner, his duties and responsibilities.

d. *Contractor.* This section defines the responsibility of the contractor in payment of costs for labor and materials, permits, taxes.

e. *Subcontractor.* This section defines a subcontractor, conditions for use of subcontractors, required listing of subcontractors, and approval thereof. All stated obligations and working relationships are given.

f. *Separate contracts.* This section gives owners rights to award separate contracts in connection with other portions of work related to the project and the responsibility each contractor has toward the other.

g. *Miscellaneous provisions.* Typical topics under this article are: law of the place, claims for damage, performance and payment, bonds, patents, arbitration, and other topics peculiar to the project.

h. *Time.* This section defines the time for beginning and completion of the work, procedure for claims in time extension. Penalties for failure to complete on time and any other liquidated damages should also be given here.

i. *Payments and completion.* This section covers all provisions relating to progress payments, certification of progress and completion, and procedure for final inspection.

j. *Protection of persons and property.* This section sets forth the responsibilities of the contractor to protect the safety of all persons who come on the job site and to protect all property.

k. *Insurance.* This section lists the types of insurance (liability, property) and amounts the contractor is required to carry. Any insurance to be carried by the owner is also given.

l. *Changes in work and correction of the work.* This section details the procedure for changing any part of the work and the way in which the cost of such changes will be determined. When any part of the work does not meet the inspection of the owner or his representative the contractor is required to promptly correct the work at no cost to the owner.

m. *Termination.* Justifiable reasons for terminating the work of the contract by either the owner or contractor are defined in this section.

Many organizations that let construction contracts have, over the years, developed a detailed text of the General Conditions. This same text may be used without any appreciable change from project to project.

The American Institute of Architects has a set of Contract Documents consisting of Agreement forms and General Conditions that can be purchased. These can be used as guidelines or in total as an engineer or architect may wish. The General Conditions are too extensive to be printed in this book. Reference 1 has these AIA contract documents printed in the Appendix.

It is considered prudent to have the contractor and contracting officer initial each page of all contract documents at the time of signing of the contract. This would include initialing all drawings and specifications. This may be a formidable task on a large project. It does, however, prevent either party from claiming a change has been made without the knowledge of the other party. The initial will certify in court that both parties have seen the page and a plea of ignorance of its contents cannot be made.

Drawings may have changes inserted after the contract work has begun. All changes should be noted in a special place on the drawing and initialed by engineer and contractor. Each revision will carry a revision number and date. At the time of receiving a revised drawing or change order the contractor will give notice within a given time period to the owner's representative by letter if he will not perform the change in work without extra compensation. In order to have a justifiable claim for extra compensation, the contractor must be able to prove to the owner and engineer that the change will require extra labor and/or additional material.

D. DRAWINGS

The drawings and specifications are the heart of the construction contract. The drawings tell the contractor, in detail, what work he needs to perform. As the old cliche says, "a picture is worth a thousand words," a drawing may be worth several thousand. Drawings are the communication medium for engineers/architects and constructors. It would be almost impossible, even on a small construction project, to give in words all the required details of the construction. Drawings should be clear and specific as to the scope of the work.

The drawings are prepared by the engineer/architect and are called the *design drawings*. After the bidding and at the time of signing of the contract these design drawings become *contract drawings*. Lump sum contracts must have specific drawings and be comprehensive in content.

There should be no question as to the extent of work. Anything left to the discretion of the contractor can be a source of trouble, and possibly legal conflict at a later date. There is no excuse on the part of the engineer/architect for inadequate or erroneous drawings. Poorly prepared design drawings can lead and have led to legal claims against the design engineer/architect and/or owner. Engineering and architecture students should take every opportunity to study design drawings to develop a facility for reading such and to also observe the contents, arrangement, and degree of detail.

During the course of the construction there may be differences of opinion between the owner or his representative and the contractor regarding the scope of the work or on some detail of the construction. Reference to the drawings should resolve the argument. If they do not, then the drawings have been improperly or inadequately prepared. In several decades past, drawings were almost looked upon as works of art. The original drawings of the Brooklyn Bridge fit this category of art. Today the emphasis is not on art but on clarity and accuracy.

Today there is hardware available for the preparation of some types of drawings by computer. However, the skilled, experienced draftsman is still a very important cog in the machinery of engineered construction. The design engineer or architect has to have the ability to translate the design calculations through the draftsmen to the drawings. Design calculations are of minor worth without the ability to transmit that information to a proper drawing that tells the contractor the scope of the work.

The design drawings may be few in number or very extensive. On large projects they will be organized into sections. For instance, the drawings for an office building will be organized into architectural drawings showing all dimensions of the building including door and window locations. Wall, floor, and ceiling surfaces as well as types of doors and windows will be shown on the architectural drawings. Cabinet work, if any, will also require drawings to be included in this first set.

The second set will be the structural drawings prepared by the structural engineer. These will give details and dimensions of all the structural elements such as footings, beams, girders, columns, and floor slabs. These structural elements will most likely be made of either reinforced concrete, structural steel, or a combination of both. Drawings of reinforced concrete elements will show all dimensions and location, size, and quality of reinforcing steel. The drawings for structural steel elements must show the type of element, length, dimensions, and all details such as details of fasteners and connections of the various elements. The drawings should specify the type or grade of material. This information should also be given in complete detail in the material specifications.

The third set of drawings in a building project will be the mechanical drawings giving all the plans for supplying and installing all the mechanical equipment such as plumbing, heating, and air conditioning. Any special equipment such as temperature control rooms, elevators, etc., will also be included in the mechanical or electrical drawings.

The next set of drawings will specify the electrical installation. The electrical subcontractor will work with these drawings. All light fixtures and electrical motors will be shown or specified on these drawings. The electrical engineers will specify much of the electrical equipment by symbols on the drawings. Electrical drawings will not show a "picture drawing" of equipment but only a symbol representing the item and location. Careful coordination between mechanical and electrical drawings is very necessary.

The remainder of the drawings will show special additional work such as landscaping, sidewalks, driveways, etc. The design professional must be sure he has covered all the work to be done that the owner expects. Many conferences and much communication must take place between the owner and the designer during the design phase of the project. The designer will have many choices during the course of preparing a design. He will seek the desires of the owner on many choices yet on those of a more technical nature, he may make a unilateral choice. Some selections the owner may defer to the designer since he will have the background to make better judgment.

The contractor's tradesmen will most likely not work from the design drawings. Various trades (subcontractors) will prepare their own *shop drawings*. These drawings will have detailed dimensions so that the element shown on the drawings can be made—usually in the shop. For instance, shop drawings will be made of all structural steel. These drawings will show the overall dimensions of each piece of structural steel, the type and size of the structural shape and the dimensional location of all bolt holes or weld sizes. Connections will also be shown in detail.

It is necessary for the design engineers to check the shop drawings to be sure that the correct material sizes are being specified as well as correct arrangement of materials. However, the detailed correctness of shop drawings are the responsibility of the contractor—not the engineer or architect. To be sure this is understood by the contractor, a statement in the General Conditions as well as the approval stamp that the design professional places on the shop drawings should read "correctness of all shop drawings is the responsibility of the contractor; approval by the engineer/architect does not relieve the contractor of this responsibility."

At least one record set each of the contract drawings should be preserved by the owner, E/A, and contractor. Since some liability claims have developed many years after completion of a project, it is very necessary that drawings be available for any future reference.

Cost-plus contracts will not require the complete design drawings at the time of contract. Only sufficient drawings to show the scope of work will be absolutely necessary to execute a contract and fix the contractors fee. However, design drawings sufficient to start the work will be necessary at the starting date. Other drawings must be available in time for the contractor to plan the work ahead. The owner and his design professionals should not delay the contractor, since this could result in claims against the owner for additional fee.

A careful listing of all design drawings should be shown in the contract and, of course, indexed on the first page beyond the title page of the set of drawings. Omission of a drawing will then be detected. It will then be the responsibility of the contractor to perform all the work as given by all the drawings for which he will receive his bid price.

Where required, drawings giving additional detail can be transmitted to the contractor after the work is in progress by a job instruction form. If the added drawings do not increase the scope of the work but only clarify some detail, then the contractor is not entitled to additional compensation. However, if work is added or deleted, then the owner, engineer/architect, and the contractor have to agree upon a change in contract price. The mechanics of this is given in a following section of this chapter entitled Payment. The General Conditions of the contract should state that the scope of the work may be altered after work has commenced by a *change order* issued by the engineer/architect. It is a rare project that will not have several change orders during the period of construction. Development of new materials or new equipment can take place between the time period of the completion of the design and the finish of the work. The owner's requirements may change and unforeseen circumstances may develop during the period of the project. Neither the owner nor the contractor should receive a "bonanza" or a financial hardship when a change becomes necessary. Each should be fair and ethical with the other. The engineer/architect can be of valuable influence in helping the two parties arrive at a fair price for any change.

E. SPECIFICATIONS

The specifications are part of the contract documents and describe in words the materials and workmanship required in the project. The specifications must be written in clear, precise language that is standard vocabulary of the design profession and trades. The contractor as well as the owner will be bound by the specifications. The contractor must supply the quality of material as specified but no more. The specifications should be technically accurate and call for the required material and workmanship.

For projects of large magnitude the specifications will be written by engineers or architects who are especially skilled and experienced in specification writing. Specification writers will work closely with the design engineers to make certain the proper material and workmanship is specified. In the design of small projects the design engineer will probably write the specification. Standard specifications such as ASTM, AISC, SAE, etc., may be made part of the total project specification by reference to part or the entire standard specification. Since these standards change with time, reference should be made to the latest edition. All contractors should have ready access of these standard specifications.

Completeness and clarity of the specifications are of prime importance. Specifications are not written as a literary effort. Use of synonyms is to be avoided. The same term can and should be repeated as often as necessary. Some key terms may have to be defined. Some terminology used by the construction mechanics may not be technically correct and use of these words in the context as used by the mechanics should be avoided. Two examples are the words "cement" and "iron." The technically correct terms should be concrete and steel when such is to be used. Cement and iron are only components of concrete and steel.

Specifications should be complete but as limited as possible in wording. Unnecessary repetition should be avoided. However, thousands of dollars have been lost by both owners and subcontractors by the omission of a single word in a specification. Sometimes writers will intend for an adjective that precedes a noun to also apply to a second noun following thereafter in a sentence. Courts have ruled that the adjective did not apply to the second noun also. Past contracts have ended in litigation because there was no agreement on the meaning of a word or words. When there is a difference of opinion on the interpretation of words in a contract the court will attempt to establish the accepted meaning by the profession. This will require testimony of expert witnesses. In all instances the court will attempt to determine the intent of the contract and not to rely on a dictionary definition of a word or words. The judge will investigate the surrounding circumstances in the development of the contract and any statements the parties may have made in the contract negotiation. Custom and usage will play an important part in establishing the meaning of a disputed part of a contract.

Where there is an unresolved conflict in the meaning of a part of a contract, the guide to court decisions is against the party who was responsible for writing the contract and who is responsible for the ambiguity. The party who was not sufficiently careful in his writing is penalized for his lack of care. On most products the owner will rely on the engineer to specify the material and workmanship that he believes to be best for each particular item to be supplied or constructed. For example, there are several grades of reinforcing steel that can be used in a concrete structure. Each grade has a different strength and also a different cost. The structural designer must select the grade that will result in the optimum design based upon cost. Some engineers may feel they should have complete discretionary power with regard to the specification since in the final product rests his reputation and his future. His reputation will rest with the quality of his work; however, he will show bad judgment if he does not listen to the wishes of the owner before making a decision. When he cannot agree with the owner's desire he should educate the owner in the reasons for the selection preferred by the engineer.

The design of many projects will call for the use of standard products. Most standard "off-the-shelf" products will be of variable cost and quality. The engineer/architect will be required to specify a grade of product. This may be simply done by naming a particular brand and model number. On government projects this naming of a product from a single manufacturer will likely be illegal and could result in legal action from other manufacturers. This can usually be avoided by giving a brand name and then the phrase "or approved equal (or equivalent)." This then leaves the door open to the use of other brands if the manufacturers can prove their products as equivalent. On private projects the designer can specify a particular product, but unless he feels this single brand product is the only one that will meet the requirements, the "or approved equivalent" phrase may save him ill feelings from manufacturer's representatives.

Specifying a single brand will reduce the price competition for the item and thus not be in the best interests of the owner unless the particular brand is unusually superior to all other brands and is necessary to fit an existing system.

A weakness in some specifications is the use of a term that is supposed to define quality but leaves the degree of quality undefinable. An example is the phrase "all workmanship is to be first class." "First class" is undefinable and at best it can only mean a standard that would be considered acceptable by third parties. The statement does serve notice to the contractor that "shoddy" work is not acceptable. To avoid any future difficulties in the acceptance of the workmanship of the contractor, the quality of work should be defined if possible. If such a definition of quality is undefinable, then the next best policy is to meet with the contractor before each phase of the work and reach an agreement as to the workmanship that is to be expected. The owner or his representative will be more successful in demanding a high quality of workmanship before the work starts than by rejecting work that has been done and attempting to require the contractor to do the work over again. It is just human nature to resist doing again what one has just completed because it doesn't meet with another's approval.

A very important concept of construction practice that the novice engineer should learn is that a contract (especially the specifications) can specify what the finished product should be but cannot specify exactly how the work is to be performed. If the specifications dictate the procedure the contractor is to follow, the contractor is not liable for the results. Contractors should be given as much freedom as possible in their operations. This will lead to a lower bid price, better working relations, and usually an earlier completion date.

However, in some types of work the design engineer may want to approve the method of operation before the contractor begins the work. This is usually the case with contracts for bridge construction. The designer should leave the door open for the contractor to erect the structure in the cheapest manner and in the way he believes is best. However, care must be taken to insure that the structure is structurally capable of being erected in the manner planned by the contractor. A specification that carried the statement, "The method of erection must be approved by the design engineer," prevented a contractor from erecting some long prestressed concrete beams by a method that would have cracked the beams. Submission to the design engineer of drawings showing the erection method prevented a costly mistake.

Itemizing of work to be performed in a contract can lead to problems. A general overall statement of the work to be done is better than a detailed itemizing of the tasks involved. A rule of contract law is that the express mention of one thing implies the exclusion of another not mentioned. Care must be taken in the wording. For example, a contract may state that "the contractor is to fabricate and erect on the site designated an industrial building as shown on the drawings." Since there is a necessary item of activity, namely, the shipping of the structural steel from the fabricating plant to the erection site, that was not mentioned, a claim over and above the contract for the cost of shipping steel could be made by the contractor. This possibility exists unless the contractor did include this in his bid price. Past history has shown that some contractors have looked for such loopholes in contracts where they observed an item that was missing in the contract. Leaving this out of the bid may make them the low bidder (if other contractors did not do the same). After signing the contract, an extra is claimed for the missing item or unmentioned activity.

All change orders to the contract that pertain to changes in the specifications should have the new specification item clearly detailed and all parties to the contract should sign the change order and a copy of the signed change order should be kept by all parties to the contract.

F. PAYMENT

The General Conditions of the contract should clearly specify how and when the contractor is to be paid. Only on small projects will the contractor be paid the total contract price at the final completion of the work. For a medium sized project the contractor may be paid after each increment (say 5 percent increments) of the work is completed. On projects that will be a year or longer in duration payments may come monthly. The amount of the monthly payments can be specified in several ways.

1. The engineer/architect may make an evaluation of the amount of the project accomplished in the month, say five percent. The owner will then pay the contractor five percent of the contract price less a retained amount of about five to ten percent of the payment. The retainer will be paid at the end of the contract and will insure that the contractor will complete the job in all details.

2. In a contract that is cost-plus the contractor will submit bills of materials and labor vouchers at the end of each month. The owner can pay these costs directly or pay the contractor. The contractor will then be paid a percentage of his fee in addition. When the government is the owner, the material bills may be paid directly to the suppliers to avoid the payment of sales and possibly some excise taxes.

If the contractor is required to finance the project over a period of time then he will have to seek financial backing and the interest payments will then be included in the bid price. Since the owner will most likely have to obtain money for the project by loans or bonding, there is no need to require the contractor to borrow money to finance more than a small cost of his operation.

If the engineer/architect is required by his contract with the owner to make an evaluation of the progress of work at the end of each month then the cost of this must be included in the engineer/architect fee. This can amount to several man-days of work on a large project.

G. INSPECTION

The inspection phase of the project is an important part of the total project. Inspection is necessary to insure that the work is accomplished in conformance with the plans and specifications. The majority of contractors are dedicated to doing the work as required, supplying the specified materials and doing quality work. A contractor who doesn't follow this principle may find his reputation is such that he is no longer on the list of approved contractors. The temptation to "cut corners" and perform the minimum work necessary is real, especially when the accounting shows that the bid price was on the low side.

The inspection of the work is provided by the owner, engineer/architect, or by an independent inspection firm. The inspection contract is paid by the owner and not by the contractor. In some past projects, inspection work such as casting and testing concrete cylinders, and nondestructive welding tests have been part of the contractor's obligation. In such cases the contractor becomes his own inspector. This does not appear to be the most desirable procedure.

The engineer/architect's contract should detail what inspection is required and the contractor should also have details of the required inspection. The contractor may be required to notify the owner or inspector when certain materials have arrived at the job site so inspection can be made before installation. The inspection may require the conducting of certain tests. If this is the case, and the contractor's work force is involved in the running of the tests, then this work should be clearly specified in the contract. If not, then the contactor will have a claim for extra compensation.

The inspection must be prompt so as to cause no delay to the contractor. Some work, such as reinforcing bars, may be covered up by the contractor before inspection takes place if the inspector is not constantly present at the work site. On large projects the engineering firm will have a resident engineer at the site to supervise inspection and available for questions by the contractor.

Regardless of the work being passed by the inspector the contractor is still responsible for his work. Deficiencies found later, even though inspected, would be the responsibility of the contractor.

Good inspection requires careful, conscientious work and good judgment. An unqualified, overzealous, or sloppy inspector can create many construction problems and a poor relationship with the contractor's employees. An inspector is subject to the blandishments of unethical contractors and material suppliers. The inspector should not try to "make life hard" for the contractor yet he is duty bound to see that the specifications are followed and that the owner receives the facility as specified on the plans and specifications. The author's experience has been that when the inspector indicated early in the project that he would be reasonable yet would make very sure that *all* work was done as specified, and that he was serving the owner, the entire project went smoothly and there was less friction and very few rejections of material or workmanship. A too lenient inspector, at the beginning of a project, may find difficulty throughout the entire project.

Work that does not meet the specifications should immediately be brought to the attention of the contractor's foreman and also reported to the resident engineer. A record should also be made of the deficiency. The inspector should keep a log book recording work inspected, those present, and the evaluation of the work. Such a log will be valuable if any legal claims are later made against any of the parties to the contract. The inspector *never* gives orders to the contractor or any workman. He only indicates if the work is or is not satisfactory.

If there is a deficiency or error in the drawings or specification this should be brought immediately to the attention of the engineer/architect so that a change order can be issued. The inspector does not make any changes to the work as specified by the drawings and specifications.

Rejection of work performed by the contractor when he has carefully followed the specifications can lead to extra cost to the owner. At times the finished product is not what the owner or engineer/architect desired. The question then arises as to whether the specifications or the workmanship is at fault. If the work has been closely followed by the inspector, it should then be possible to determine if the work procedure and materials were according to the specifications. If so, then the specifications may be at fault and any remedial work would require additional cost to the owner. Close inspection by a qualified inspector can save disputes and possible court action.

When the work is not proceeding as specified, the inspector should stop the work and immediately notify his supervisor. Orders to stop the work should be given to the contractor's superintendent who is resident at the job site. Workmen are never given orders by the inspector or engineer/architect. Workmen only receive orders from their foreman. Failure to follow this important procedure can result in additional claims by the contractor, as well as develop friction between the contractor and the owner's representatives.

H. FINAL INSPECTION AND ACCEPTANCE

As a project is nearing completion there will be a complete inspection. This will be called by the contractor upon approval of the engineer. Except for minor "touchup" and final cleanup all major activity should be completed by the time of this inspection. Just prior to this meeting the resident engineer should have his inspectors make a list of all unfinished or unacceptable work. The resident engineer and the contractor's superintendent will then conduct the inspection with any other representatives the owner and contractor desire. Subcontractor's representatives will likely be present when inspection is made of their area of work.

On large projects the final inspection may take several days. When supply and installation of operating equipment is part of the contract, this equipment will be tested during the inspection. Heating and air conditioning systems must be thoroughly tested and inspected. It may be necessary to delay such tests until the proper climatic conditions exist.

It may be the case that particular subcontract work will have a final inspection some time prior to completion of the entire project. It may be desirable to do this before the subcontractor moves off the job. However, the prime contractor is responsible for this inspected work until the end of the project. The final inspection will review this already accepted work to see that no damage or loss has taken place since the previous inspection. A subcontractor can be called back on the project at any time until the entire work is completed and turned over to the owner unless his contract states otherwise. However, it is a better policy to make a final inspection of the subcontractor's work as soon as it is completed and then release him if all is satisfactory.

During the final inspection a list will be kept of items of work that are not satisfactory. Upon completion of the inspection the owner or his representative will sign the list of deficiencies and the contractor will do likewise and each will retain a copy. When the contractor has corrected all deficiencies and he is satisfied all the work as directed in the plans and specifications has been performed he will then notify the owner of such. A second checking of the project will then be conducted with specific inspection of those items on the deficiency list. The engineer's inspectors should also check all parts of the project to see that nothing has been damaged in the interim period.

If all work is satisfactory to the owner, then the contractor is given a statement of completion. The owner then should receive a letter from the contractor stating that the project is fully released from all claims and liens (*see* Chapter 11, Section I). The sureties (*see* Chapter 10) should also give written consent to the owner to make final payment to the contractor. The sureties will not give this consent unless all bills and labor have been paid. A statement should be in the contract that the final payment does not release the contractor or his sureties from their obligations under the contract or under the payment and performance bonds.

Final payment does not release the contractor from repairing or replacing any faulty work or material. The courts have required contractors to come back to the project and remedy any faulty work or equipment. The contract may contain guarantee requirements of varying periods depending upon the item. Difficulty can arise with a guarantee when the contractor claims the equipment has been misused by the owner.

Sometimes the owner will occupy and use the facilities before completion. This should be an agreed-upon arrangement, especially when the contractor has not completed all items by the contract completion date. This occupancy may be necessary for rental property where the owner has signed leases with a guaranteed starting date. Before occupancy a complete inspection should be made so that no claims will be made against the contractor for damage done during occupancy. The owner will also be protectd from claims by the contractor that unsatisfactory work was due to damage during early occupancy.

The doctrine of substantial completion or performance has been established by the courts to give release to contractors when all the work is complete except for minor details. A 1959 court decision gave the opinion, "There is substantial performance when the builder, in good faith, has intended to perform his part of the contract and has done so in the sense that the building is substantially what is provided for, and there are no omissions or deviations from the general plan that cannot be remedied without difficulty." This doctrine of substantial performance permits the following events to take place:

1. The period for assessing liquidated damages (*See* Chapter 11, Section B) ends.
2. The owner can occupy the facility.
3. Final payment can be made to the contractor except for the amount that it would take to complete the work plus a reasonable amount of money to assure the contractor finishing the work in an expeditious manner.
4. Responsibility for the facility shifts to the owner.
5. The guarantee period starts.

The date of substantial completion is usually recommended by the engineer/architect in consultation with the owner. This date should be fair to all parties.

It is almost always to the contractor's advantage to expedite the work as rapidly as possible, especially near the end of a project. Not only getting the job done and receiving final payment enhances the reputation of the contractor, but in a world of high interest rates, money can be lost by failure to move toward rapid completion. Labor should not be allowed to drag near the end of a project. This has to be supervised very carefully on cost-plus contracts.

I. AVOIDANCE OF PROBLEMS IN CONSTRUCTION CONTRACTS

Many different legal problems can develop in a construction contract. Legal history is full of such problems. There is not enough space in any book to even summarize the multitude of problems that may be encountered. A few procedures will be given here that if followed will probably prevent a number of mistakes that are common to construction contracts. Avoidance of legal pitfalls is much more prudent than trying to resolve the problems after they occur. Chapter 11 describes a number of construction contract problems and how such problems have commonly been resolved in the courts. The following listing enumerates procedural steps that may help prevent legally entangling problems.

1. Public contracts especially, but also private contracts, should be uniformly drafted and rigidly enforced. Many problems arise from a poor understanding of the contract by the contractors, as well as the attitude, due to loose control, that it is very easy to "get away" with not conforming with all items in the specification.

2. Contracts of public agencies should be as uniform and as consistent from one job to another as possible. Improvement should always be sought after, but change for just the sake of it can lead a contractor to a pitfall. Pre-bidding conferences should serve notice to contractors of meaningful changes in contract clauses and specifications from what may have been standard procedures in past contracts.

3. All bidders should utilize a specification checklist to make sure they have read all provisions and have included all specified items in their prepared bid.

4. The design professional should explain bidding and contractual procedures in detail to all owner representatives before the project goes to bid. The manner and cost of making changes in a contract after it is signed should especially be covered. This is especially necessary when the contracting agency is a governmental or quasi-governmental administrative body. Many such inexperienced people believe that no matter how many changes are made in a lump sum contract the total price remains the same. This book or a similar one should be required reading for every newly elected public office holder who will be dealing with contracts of any kind.

5. Design professionals should acknowledge any errors or mistakes in the plans or specifications as soon as recognized and to correct them as quickly as possible. Uncorrected errors can sometimes "snowball."

6. All changes from the plans and specifications of the contract should *always* be in writing and copies quickly transmitted to contractor and owner. These written changes are titled Job Instuctions or Change Orders.

7. An ethical, businesslike relationship between all parties to the contract should be developed on any project. The project should be devoid

of any indications that the owner or contractor is attempting or will attempt to get something for nothing. The component of each party to a contract should conscientiously convey confidence in honesty and fairness to the other parties.

REFERENCE

1. J. Sweet, *Legal Aspects of Architecture, Engineering and the Construction Process.* St. Paul, Minn.: West Publishing Co., 1970.

10. PROPOSALS, BONDING, AND SUBCONTRACTS

A. PROPOSALS

All contracts start with a proposal. It is the beginning of a mutual agreement between parties. The proposal is a promise from one party to another to perform a certain act in return for a consideration. Acceptance of a proposal is a basis for a contract. In order for a legal contract to be developed from a proposal, all the elements of the proposal must meet the legal requirements of a contract, as set forth in Chapter 7. The proposal must have subject matter that is legal and specific; there must be a proper consideration; the proposal must be offered by a competent party to another competent party.

The preparation of a proposal is a very important matter, regardless of the type of action involved in the proposal. A proposal that has elements of illegality can lead to criminal indictment. People in all types of business activity become involved in offering proposals or being offered a proposal of one kind or another.

In Chapter 9, the details of preparing a bid proposal by a construction contractor is outlined. In order for the owner to evaluate all bids properly, the owner should make certain that all the proposals from the several bidders will propose doing exactly the same thing. Therefore, standard forms are utilized in construction contracts and all bidders use the same drawings and specifications.

Engineers, in specific technical activities, may have to evaluate proposals that are not the same in content. When the proposal is a construction bid based on a complete set of drawings and specifications, then the subject matter is the same for all bidders. However, if the subject matter of the proposal is to perform a body of research, to make a feasibility report, or prepare a design of a given facility, there is no definite specified range or detail of activity. Several proposals may in general or in some specifics be similar but the amount of work and the expertise involved in accomplishing the project can be greatly different among several organizations. Therefore, when several proposals are evaluated and one is selected primarily on the basis of lowest bid price, there is a false value judgment. Bid price can be the basis for selection only when the specified work to be performed is exactly the same for all those submitting proposals. Some private owners have believed that even on a construction project that is clearly and definitely defined by a set of drawings and specifications, there can be such a difference of performance that they consider only one proposal. The single invited proposal is from a trusted contractor whose past record for doing quality, on-time work is excellent. In such procedures, the price is negotiated.

For construction contracts, the proposal is usually required to be submitted in a particular format. This is discussed in detail in Chapter 9, Section B. In such a business activity, it is very important that the proposal conform to the required format. Even a minor deviation can be cause for rejection of a proposal.

In other types of business activity, there may be no special form or procedure. Many times a business entity will present an unsolicited proposal to another party. This proposal can be the result of the submitting party sensing a need by the receiving party of a new product, investigation, study report, etc. An example might be where manufacturer A is producing an assembled product. Manufacturer B observes that one component of the assembled product can be improved upon by a different design and a different material. Manufacturer B then prepares and submits a proposal to A to redesign, test, and supply several components. The proposal by B will have to contain convincing data that they can produce a better product. If A is convinced of such, a contract may follow.

A recent case involved a technical development and manufacturing firm which submitted a proposal to a military branch of the Federal Government to design, test, and manufacture several field test items of a piece of military equipment. The item, which was already being used, was of steel and the proposal was to replace this with a substitute manufactured from spun filaments. The great advantage of the substitute was its considerable lighter weight and corrosion resistance. The military accepted the worth of the proposal; however, because the activity would involve a contract with the Federal Government, it was necessary to publicly advertize for proposals from other companies. Several other proposals to perform essentially the same activity were received.

B. ACCEPTANCE AND REJECTION OF PROPOSALS

Before a proposal can end in a contract, it has to be accepted. Acceptance can take several forms. If there is an acceptance without any qualifications or changes in the proposal, then a contract has been made. This acceptance can range from a simple verbal acceptance to a very formal written contract such as for a construction project. Once a proposal has been accepted, it cannot be withdrawn without a breach of contract. A breach of contract may lead to damage claims by the accepting party. Bid bonds are submitted in construction contracts to cover damage claims if the party submitting the proposal withdraws from the contract before beginning work.

A proposal can be rejected by the party receiving such without any claim against the party. It is obvious that there is no contract and thus no breach if a proposal is rejected. The rejection can be in the form of no notice of acceptance (a rather rude procedure) or a verbal or written notice of rejection. A preferable procedure in either rejecting or accepting a proposal is a written notice. A receiver of proposals has the legal right to reject any or all proposals. It may appear somewhat unfair that one can go to considerable expense in preparing a proposal (a bid in a construction project) and have the recipient (owner in a construction project) reject all proposals. Nevertheless, this is legally permissible, and to operate otherwise would introduce very complicated problems of contract law.

Some proposal invitations have solicited design competition among design professionals. Such competitions have taken place among architects wherein the facility to be designed is very noteworthy and the aesthetic qualities are very important. When the competition is open to anyone then the rejected designers receive no reimbursement for their efforts. If the competition is open only to a selected few then the losers may or may not receive modest funds to cover some of the cost of developing their proposals.

Design competition in traditional civil engineering structures is not common in the United States but not unusual in other countries. However, in Europe the design and construction is commonly linked together in one proposal.

Proposals for large design projects may involve considerable manhours of effort in preparation of the proposal document and possible travel costs of several professionals to make the presentation of the proposal in person. The selection and acceptance of the winning proposal by the owner can involve many hours of study by several of the owner's representatives.

When a proposal is returned for modifications then the party submitting the proposal can accept or reject the suggested modifications or changes. No acceptance of the proposal in its original form voids the proposal and a resubmittal of a proposal is legally the same as a new proposal. Proposals for construction work or the supplying of materials usually carry a time limit. It is good policy to have a terminal date after which the proposal is null and void unless formal acceptance has been received. Conditions, costs, etc., will change with time and the party making the proposal needs to provide protection from future changes.

When all the proposals for a construction project are rejected as being too high, then the project should be modified to reduce the cost before new proposals are accepted. The act of "bid shopping," wherein the owner rejects all bids because the lowest bid is too high and then accepts a second round of proposals without a reduction in the project is considered bad practice.

The practice of meeting with the lowest bidder to negotiate a lower bid, with or without project modification, has been done by private owners when the amount of available funds are less than the lowest bid. If this procedure is followed by a government organization, it can lead to legal complications. The other bidders could claim they did not have the opportunity to bid on the modified project.

C. BONDS

There are a number of types of bonds that may be required to be part of the proposal on construction contracts. A construction bond is a protection to the owner from suffering a financial loss due to the fault of the contractor. These types of bonds are called surety bonds. They are a form of insurance that the contractor has to provide for the owner. The purpose, use, and legal rights of all parties under different types of bonds should be understood by anyone involved in construction contracts. The basic principles are presented herein.

In 1894, the Heard Act prescribed the use of a contract bond in federal construction contracts and in 1935, the Miller Act separated the single bond into one assuring performance of a contract and a second bond covering default on the part of the contractor in paying his bills for labor and materials.

A construction bond is a guaranteed payment of money to the owner by the contractor or some separate organization which has taken on the obligation of the payment to the owner when the contractor defaults. The organization which assumes the financial obligation of the payment of the bond principal is called the *surety* or the *underwriter*. The surety gives financial backing to a contractor for a given, designated amount of money. For this financial guarantee, the contractor pays the surety a given amount of money called a premium. The premium will be a given percentage of the face value of the bond. The percentage may vary among contractors. A well established contractor with a record of successful completion of projects similar to the subject project will have a low premium rate and thus some advantage in the bidding. A contractor who defaults on a previous job may have a difficult time obtaining a bond for a subsequent project. The bond premium will be passed on to the owner in the bid price.

The surety is a financial firm which furnishes bonds of various kinds to business organizations. Bonds to contractors are an important part of such a firm's business. The bond is a legal document that is given to the contractor, and is then passed on to the owner. It guarantees the payment of the face value of the bond if the contractor defaults in his contractual obligations to the owner. The financial standing and worthiness of the surety is of considerable importance to the owner. The owner should require that the surety be approved before the submission of the bid. The types of bonds that may be required in a construction contract and the purpose and amounts are described below.

1. *Bid Bond.* The function of the bid bond is to insure (or at least provide considerable assurance) that the contractor selected by the owner will sign the contract and furnish any additional required bonds. This bond must be part of the bid proposal submitted at the time of bidding. If the selected contractor does not sign the contract then the bond is forfeited to the owner. The amount of the bid bond is selected by the owner prior to submission of bids. A typical amount of the bid bond is five to ten percent of the bid price. The average cost of a bid bond will be about five percent of the face value of the bond. The higher the value, the less the percentage. The cost of the bond will have to be included in the overhead calculated by the contractor for the job. Thus, the owner pays for the cost of the bond in the contract price. Therefore, the owner should not require a higher bond than he thinks is necessary for his protection from a defaulting contractor.

An example of how the bond works is shown where bids are taken and the low bidder submits the price of $245,000 to do the work. The bid bond was $20,000. Before he signs the contract, he decides he does not want to do the job since costs have gone up and he has had some key personnel quit. The second bidder bid $260,000. Because of default of the low bidder, it will cost the owner $15,000 more for the project. The owner can then claim $15,000 of the bid bond. This amount is then paid to the owner by the surety. The owner cannot collect more than the $15,000 loss even though the bond would cover up to a $20,000 loss. An owner cannot make money because of default of a bidder. He can only cover his losses. The low bidder loses his premium (he does not get the job so the premium is not covered) and he loses his reputation with surety companies. On future jobs, his bid bond premium will be higher, if he can buy a bond. If the second low bid was $295,000, the owner could only collect $20,000 from the surety and the owner would suffer the extra $10,000 loss.

Some proposals state that upon failure of the selected bidder to enter into a contract the bid bond will be forfeited as *liquidated damages* (*see* Chapter 11, Section B). Liquidated damages can be more than just the difference between the low and second bid. It can also include costs required to execute a contract with the second low bidder. Part of these costs could include loss of time.

> 2. *Performance Bond.* The purpose of the performance bond is to assure that the owner will receive the constructed facility for the contract price. The performance bond as purchased from an approved surety is required at the time of the signing of the contract. The face value of the performance bond is whatever the owner desires. The owner must be mindful that the bond premium will be part of the contract bid. The required amounts of all bonds are part of the information to all prospective bidders. A typical value of a performance bond is 100 percent of the bid price. This amount will most likely fully protect the owner from the contractor's failure to complete the work as required under the provisions of the contract.

The terms of the contract or bond require the surety to promptly remedy the contractor's failure to perform. Surety companies will usually engage a second contractor to take over the work upon the default of the original contractor. They will enter into a contract with the substitute contractor. The owner will then make payments to the surety upon progress of the work. Most likely, the surety will lose money because of the default, since the second contractor will usually require more money to do the unfinished work than was allocated for the residual portion in the original contract. Suppose an owner signed a contract with a builder to erect a building for $2.4 million dollars and required a 100 percent performance bond. After the building is about one-quarter complete, the contractor gets into financial difficulty and declares bankruptcy. He then defaults on his contract. The owner has paid the contractor $450,000 in partial payments for the work he has done. The surety then takes over the project and calls for bids to complete the work. There are unpaid material and labor claims against the work that has already been constructed to the amount of $30,000. The surety accepts bids for the unfinished portion and the low bid is $2.1 million. Since the owner has a performance bond for one hundred percent ($2.4 million) and some work is already complete, he will not likely have to pay more than $2.4 million for the building. He will have to pay the surety or another contractor the difference between the $2.4 million and the $0.45 million he has already paid or $1.95 million. The default of the first contractor will cost the surety $2,100,000 plus $30,000 less $1,950,000 = $180,000. If the surety could obtain a lower bid than the $2.1 million, it would cost him less even down to no cost at all if he could obtain a sufficiently low bid.

If an owner does not call for a sufficiently large performance bond, it is possible that the bond would not cover the cost of completing the work. Irrespective of the circumstances or the amount of the performance bond, the owner cannot make money due to a default. He can at best only obtain the building he contracted for at the contract price.

Defaulting contractors will find it hard to obtain a performance bond on any future jobs. Surety companies keep records of contractor performance.

If an owner fails to require a bond sufficiently large enough to cover the cost of finishing the project, he may take legal action against the defaulting contractor. However, the contractor may not have sufficient assets to cover the loss.

3. *Payment Bond.* These types of surety bonds are to protect the owner against failure of the contractor to pay for the materials and labor used on the project. Suppliers of material and labor have a claim not only against the contractor but the contracted facility as well. If the contractor cannot or does not pay his bills, the creditors

can place a lien against the project in which the materials or labor was used. The owner is then required to pay the bills in order to remove the lien. Liens will be described in more detail later in this chapter.

The payment bond will require the surety to meet the unpaid bills up to the amount of the payment bond. The amount of the payment bond can be whatever the owner requires. The usual amount is between fifty and one hundred percent of the bid price. With a one hundred percent performance bond, it may not be necessary to have more than a fifty percent payment bond, especially if the owner requires proof of labor and material payments before he makes any partial payments to the contractor.

Sometimes the owner will require a combined performance and payment bond at 100 percent. If the owner keeps close scrutiny on the action of the contractor, then this may be sufficient protection. The greatest risk would be in the very early stages of the project before the contractor has performed any appreciable construction. Default at this stage is rare. Between the time of bid opening and signing of a contract, the owner should check the financial status of the low bidder.

The courts have allowed the material man or laborers to sue the surety directly for unpaid bills. However, when performance and payment bonds are together in one bond, the surety cannot be sued directly by the material man and laborers. Only the owner can sue on this type of bond. The owner can, however, be sued directly for nonpayment of bills of the contractor.

There is a responsibility of the owner and engineer/architect to the surety in the case of performance and payment bonds. When there are retainage funds in the hands of the owner and the contractor is derelict in his payment of bills, thus placing the surety in jeopardy, the owner or engineer/architect has a duty to not release these retainage funds to the contractor. The owner or his agents can be held liable to the surety for any damages. An owner had retained 10 percent of the contract price as was stated in the general conditions of the contract, such retainage to be paid to the contractor upon total performance and also upon showing proof of payment of all bills for material and labor. The owner upon certification by the eningeer, without requiring proof of payment of all bills, paid the 10 percent retainage to the contractor. It was later discovered that there were unpaid bills resulting in claims against the surety.

The courts ruled that the surety was liable for the unpaid bills less the 10 percent retainage. The owner and engineer were liable for the unpaid bills up to the amount of the retainage since they had not performed their contract obligations as stipulated in the contract documents and had thus failed to protect the surety. The owner and his agents must comply with all provisions of the contract documents regarding payment procedures to the contractor. Failure to do so can be costly to the owner and his agents.

4. *Maintenance Bond.* These types of bonds are for the purpose of protecting the owner from faulty materials or workmanship, the results of which appear after the project has been completed, the contractor paid, and the contract closed. If problems arise after the contractor has left the job and has received full pay, it may be difficult to have him return and remedy any defects. If the owner has a maintenance bond, then the surety is responsible to see that the defective work is repaired. The surety can and will put the pressure on the contractor to correct the work. If the contractor is unable to perform the work, then the surety will have to find another contractor to make the repairs or default the bond value to the owner, who will then hire someone to do the corrective work or to do the work with his own work force.

Recalling a contractor back to the project may be a problem. In some instances the contractor may no longer be in business or may have left the state. If the contractor is a reputable business firm, he will not want the surety to forfeit the bond. If such happens, the contractor's reputation will be in jeopardy and on future projects he may find that he is either unable to obtain a surety who will sell him a bond, or he will have to pay a high premium rate for bonds.

There are two possible exclusions to the payment of a maintenance bond by a surety. The first is that the defects must be due to faulty work of the contractor: the owner must be able to show that the defects are not due to poor design or improper operation or maintenance by the owner. This may be no small task if the defects are major. The second exclusion will be a limit on the time period after contract completion in which the maintenance bond is in effect. The longer the time period, the greater will be the cost of the bond; such cost is passed on to the owner in the bid price. The time period for the maintenance bond to be in effect after completion of the contract should be as short as possible consistent with the time necessary to detect faulty workmanship or materials.

It should be kept in mind by both owner and engineer/architect that although the contractor directly pays the cost of all bonds, these charges are passed on to the owner in the bid price. The kind and amount of a bond should provide only the minimum necessary protection consistent with the type of project. Bond requirements should be selected with care.

D. SUBCONTRACTS

In today's world of specialization, it is rare to have the prime contractor perform all the work with his own construction forces. Only on very small and limited-scope projects will the contractor do all the work with his own direct-hire employees. The usual practice is to subcontract out the major portion of the work. The construction industry in the United States and most developed countries is made up principally of subcontractors who specialize in one phase of construction. A prime contractor on a large project has been referred to as a "job broker," that is, he has no direct supervision of any labor but the entire work is performed by subcontract firms. The bid preparation work of the prime contractor is then relegated to obtaining bids from subcontractors, selecting the subcontracts and combining their bids into the total bid submitted to the owner. An exception to this practice is that a general contractor must perform at least half of the work on federally aided highway projects he is awarded. The Associated General Contractors are in favor of retaining this regulation. They claim that a lessening of this rule would drive small contractors out of business.

Upon signing a contract with the owner, the prime contractor has to coordinate the work of the subcontractors, see that the work is performed in accordance with the drawings and specifications, and in general serve as the coordinator of the subcontractors. As previously stated in this book, the owner has no contract with a subcontractor—only with the prime contractor. The owner's representatives will inspect the work of the subcontractor, but all decisions with regard to the acceptability of this work must be communicated through the prime contractor.

In preparing the bid for the total project, a prime contractor has to obtain bids from the various subcontractors who are acceptable and interested in doing the work. The prime contractor is not required to accept the lowest bid. He may select a subcontractor with a higher bid if he believes it is advantageous to do so. However, this higher subcontract bid will be reflected in the total contract price and may jeopardize the possibility of the prime contractor winning the low bid.

Since the awarding of a contract is based only on the price quoted by the general contractor and not on any direct quotations from the subcontractor to the general contractor, subcontractor's price quotations are not binding on the general contractor unless he has in some manner indicated an acceptance of the subcontractor's proposal. If the general contractor makes no acceptance of any subcontractor's bid proposal until he has been declared the low bidder and awarded a contract with the owner, then he is in a strong bargaining position to seek lower bids than previously quoted by the various subcontractors, that is, to bid-shop. This process of bid-shopping has been common in the past, but in recent years steps have been taken to prohibit its practice. If the general contractor can obtain a lower bid from a subcontractor than he included in his bid to the owner, then he may enjoy more profit. The financial advantage of the second-round lower bid from the subcontractor goes entirely to the general contractor and not the owner.

In the construction industry, bid-shopping is considered (especially by subcontractors) to be taking unfair advantage of the subcontractor. However, the practice of bid-peddling by subcontractors has also been common. Bid-peddling is the act of a subcontractor going to a general contractor after he has been declared the low bidder and negotiating a bid lower than that of any other subcontractor in order to obtain the subcontract work.

The acts of bid-shopping as well as bid-peddling are questionable as ethical actions. Either act can be considered to be taking unfair advantage of other persons. In both cases, the owner does not receive any monetary advantage; in fact, there are several disadvantages to the owner by these practices. Some of the most obvious are listed below.

1. In a climate of bid-shopping, a subcontractor may quote high on his original bid, knowing that the general contractor will bid-shop. If the general contractor submits his bid on this inflated quotation, and all general contractors do likewise, then the owner will be contracting on an inflated price that is the result of the practice of bid-shopping.

2. A general contractor may expect inflated prices from the subcontractors on the first quotations, and as a result submit a lower price in his bid to the owner. Upon receiving the contract he may, to his dismay, find he cannot obtain a lower price from the sub-

contractors. A sudden up-turn in jobs available, material price increases, labor shortages, etc., may be the reason. The general contractor is then faced with losing money on the contract. To make up for this, he may cut corners or resort to other practices that will be to the disadvantage of the owner. If his financial situation gets bad enough, a general contractor may default on the contract. This in turn will be very disruptive to the owner.

3. Some subcontractors may not submit price quotations if they believe bid-shopping will take place. Thus, the competition may be limited, resulting in a higher price to the owner or the use of less qualified subcontractors. The preparation of a bid involves much time and cost for a contractor and a subcontractor. Bid-shopping increases the risk of loss of that investment.

Actions of several kinds have been taken to prevent bid-shopping or bid-peddling. Some such actions are listed below.

1. Ethical sanctions. The Associated General Contractors of America have prohibited bid-shopping in their code of ethical conduct.
2. Some owners require the general contractor to submit the name of the selected subcontractors with their sealed bids. An owner may also require the price quotation from the selected subcontractor. The general contractor is then locked into these subcontract bids unless the subcontractor is not approved by the owner, or for some reason the owner approves a substitution.
3. Legislation has been passed in some states requiring the subcontractors to submit price quotations to the owner. The owner then prepares an approved list of the subcontract bids and gives this list to the general bidders who prepare their overall project bids from the price quotations on this list.
4. The practice of establishing "bid depositories" for subcontractor bids. Bid depositories have usually been operated by a trade association and the association collects the subcontractor bids. Such bids are then available to all general contractors at a stated time. All subcontractors are also given the list of all price quotations. They will then know if any bid-shopping takes place and they are in a stronger bargaining position.

Whatever action may be taken to restrict or eliminate bid-shopping, there is danger that such action may be challenged by the Federal Government's Antitrust Division. If any contract business involves interstate commerce, then it is subject to federal regulation under the Sherman Act and may be investigated for restraint of free and open competition. Some bid depositories' operations have been cited as restraints of trade. Any regulations against bid-shopping by trade associations may also be subject to action similar to what happened to the Codes of Ethics of ASCE, NSPE, and the AIA. Here again is where the Sherman Act and the interpretation of it by judges can come into conflict with Codes of Ethics.

Subcontractors should be knowledgeable in the business of contracts and the many possible legal implications. Subcontractors may have small staffs and the principals may have limited business and legal training or experience. As a result of this ignorance, subcontractors have signed contracts with the general contractor that have placed them in a difficult position on one or more issues. The subcontractor should be careful of the subcontract form they sign. In the past, many have been one-sided in favor of the general contractor. The Associated General Contractors and the American Institute of Architects have recently prepared standard forms for subcontracts that are fair to both general contractor and subcontractor. Some vital aspects of contracts that subcontractors should be aware of are listed below.

1. The warranty period of subcontractor work should start from the date that the subcontract work is completed and not the date of completion of the project. It may be that the subcontract work cannot be tested until other phases of the project have been completed. If so, then the start of the warranty period is delayed until the testing is completed.

2. The subcontractor should be sure his contract does not subject him to the liquidated damages of the general contract. Any liquidated damage clause in the subcontract should pertain only to the completion of the subcontract and not the general contract.

3. The payments to the subcontractor should be independent of the payment schedule of the general contract. If for any cause, not the fault of the subcontractor, the owner delays payment to the general contractor, this should not be reason for withholding payment to a subcontractor. The subcontractors should not be financing the project for the general contractor.

4. Retention of a percentage (5 to 10 percent) of a periodic payment due the subcontractor is usually withheld to ensure that the entire work will be completed by the subcontractor. This is similar to the retention on the total project. After completion of the subcontract work, there is no reason for withholding any money due the subcontractor. The subcontract should read: "Final payment shall be due 30 days after completion of the work of the subcontract." When performance and payment bonds are required of the sub-

contractor, then it is questionable whether a retention is necessary. In effect, this dual requirement is double protection and the owner's cost will probably be reduced by eliminating the bond or retention requirement.

5. A subcontractor should not sign a waiver of lien right against the general contractor.

6. The subcontract should carefully describe and detail the scope of the work required under the subcontract. The subcontractor should be very careful in preparing his bid to see that he includes prices for all work covered in the subcontract. The described work in the subcontract should fit the plans and specifications of the general contract. If such does not match, then there will be a later confrontation on the extent of work of the subcontractor and possible legal involvement.

7. The subcontractor should be bound to the general contractor with the same contract provisions as the general contractor is bound to the owner. To do otherwise leaves the general contractor in a vulnerable position. Likewise, the subcontractor should have the same rights of action and redress against the general contractor as the general contractor has against the owner. The extent to which the general contractor has the right to direct increases or decreases, or to suspend the work of the subcontract, should be of concern to the subcontractor. This should be spelled out in the subcontract and understood by both sides. Changes in the contract price and payment procedures under scope of work changes should be stated in the subcontract.

11. CONSTRUCTION
CONTRACT DETAILS

A. BIDDING ERRORS

The history of construction contract bids is replete with errors in sub-mitted bids. Error in a bid occurs when a contractor makes an arithmetic mistake in multiplication or addition, or misreads a dimension from a plan. In preparing a bid a contractor must "take off" quantities from a set of plans and then multiply these quantities by his selected unit prices. He must also estimate man-hours of labor and extend these values by labor cost per hour. The person preparing a bid must then sum all these individual costs and transfer these values from the work sheets to the bid forms. With a multitude of number operations a mistake is very possible. Bid preparation calls for independent checking of all cal-culations and posting of numbers. Misreading of plans is a human frailty.

Sometimes bid errors are not realized or detected by the bidder until after the bid has been submitted. Recent court rulings have stated that if an error is detected after deadline for submission of bid but before the bid opening the bidder can simply withdraw his bid with no legal action. However, he cannot resubmit a corrected bid. The bidder loses the opportunity to be considered for the contract.

It sometimes happens that a bid error is undetected until the bids are opened and compared. An error of omission or bad arithmetic can result in a decidedly low bid. If the low bid is considerably below all the other bids, the low bidder is very likely to make a hurried review of all his figures. If he detects an error that results in too low a bid, then he is likely to ask that his bid be withdrawn. He will have to prove that his reason for withdrawing is a gross error and not just a bad judgment of quantity of man-hours of labor or a low estimate of unit prices. A withdrawal without just cause can result in the loss of the bid bond.

There are many recorded court cases where the owner refused to allow the bidder to withdraw without loss of bid bond. Recent court cases have ruled in favor of the bidder when it was proven that the bid was substantially low and was due to an error in bid preparation. Rulings have not allowed an owner to have a financial advantage over a contractor because of an error due to human frailty. Such judgments have been made where the owner has suffered no expense due to withdrawing of the bid.

An interesting court case was concluded in Iowa in 1977. The owner, which was a city, detected an arithmetical error in one of the bids. The bidder had made an error in multiplication when he multiplied the quantity for one item in the bid by the unit price. With the correction, the price for that item was lower and thus the bid price was reduced. This reduced price made this bid the low bid. The owner thus made this correction and awarded the contract to this "corrected" low bidder.

The bidder who was previously low before the correction of the other bid brought suit claiming that the owner's action was not legal or justifiable. In the following court ligitation, a ruling in favor of the owner was made. The court ruled that the city did not alter a unit price, but just corrected a mathematical error. It declared that the plaintiff did not prove that the city acted capriciously, or applied any double standard. It noted that the written unit price and the numerical unit price were identical.

If the contractor has signed a contract and started work, then he is locked into his bid price, error or no error. It should be remembered that the basis for all litigation involving contracts is to render a judgment that is fair and equitable and not to permit one party to take advantage of another because of weakness or error common to human action. Errors or mistakes due to negligence are another matter and will be covered in the next chapter.

B. LIQUIDATED DAMAGES

Liquidated damages is a term used on contracts and contract litigation to mean payment of money to cover the damage sustained by one party to the contract by another because of the failure of the second party to fulfill all the obligations of the contract for which he is responsible. The term liquidated damages commonly refers to the payments due the owner because of breach of the contract by the contractor.

Many contract documents contain a clause giving the required completion date. A payment is usually required from the contractor to the owner if the project is not completed by the set date. This payment is established by the owner as a given number of dollars per day the project runs beyond the established completion date. Payments by the contractor to the owner because of failure to complete the work on time are liquidated damages.

Several years ago the liquidated damages clause for time overrun was a standard provision in almost all construction and the amount stipulated was an even amount, such as $100, $500, or $1000 per day, depending upon the size of the project. Several court cases in recent years contesting the payment of damages because of failure to complete a project on time have clearly established legal restraints on this contract provision. The courts have in general ruled that an owner can recover an amount of money from a contractor up to the proven damage caused by failure to complete the project on time provided the date and amount are written into the contract. The liquidated damage amount must be a reasonable actual damage to the owner. A case is an office building being constructed for rental purposes. If the building is not completed on time, then there is a loss of rental income. Some prospective renters may break their leases because their offices are not ready for occupancy. These damages can be established and a contractor is liable for such if there is a liquidated damage clause in the contract.

If the owner cannot prove damages of a reasonable, verifiable amount, then it is considered by the court whether the sum specified as liquidated damages is just a penalty for the purpose of forcing the contractor to complete the project on time. Such penalties have been declared void in contested court cases. Liquidated damages as a penalty has been upheld in court if the contract document stipulated a bonus to the contractor if he finished ahead of the specified date.

In a 1978 case in Georgia, the Georgia Department of Transportation assessed a bridge contractor for liquidated damages because a bridge was not completed on time. The contract contained a clause of $200 per day for failure to complete on time. The contractor sued the Georgia DOT to recover the $73,000 withheld by DOT. In the trial court the contractor won his case. However, the case was appealed and the appellate court reversed the ruling of the trial court. In their ruling, the appellate court reached the following conclusion: There was an injury to DOT because of the delayed completion date, but it was difficult or impossible to accurately estimate the actual damage. Therefore the $200 per day was a reasonable figure and was intended to be a sufficiently accurate estimate of damages and not an assessed penalty, and this was understood by the parties to the contract. Of course such cases may be appealed to even higher courts, and what a higher court may rule in such cases of fine points of law is difficult to prejudge.

The best policy for a contractor to follow is to complete the project on time by proper organization and construction management. If there is a liquidated damage clause in the signed contract, the owner will most likely assess such damages for a delayed completion date and the only recourse the contractor then has is to court action. Even if he wins, there will be legal costs plus great time involved in court by the contractor plus a possible loss in public relations. A "golden rule" for a contractor to follow is *get the job done on time and stay out of court.*

An interesting case in history is that a contract for a building was to have been completed by May 22, 1897 and was not completed until fifty-three days later. The contract called for liquidated damages of $10 per day for the total sum of $530. It was contested in court by the contractor and since the rental income of the building was $38 per month the owner was allowed a recovery of $67.13 instead of the $530.

Contrast the above case with a recent case in California where a contract called for a house to be completed in 180 days after start of construction. The house was to have been finished by April 1, 1973, but was not completed until August, 1973. There was a tax advantage loss by the owner because of the delayed completion date and the owner sued the builder and developer for a breach of contract and fraud. A jury awarded the owner $12,000 on the contract breach and $5,000 on the fraud claim. It also imposed punitive damages of $100,000 against the builder and $115,000 against the developer. The case was appealed and the higher court reduced the damage claim to $8,400 (the actual damage) and said the fraud claim duplicated the damage claim. The higher court upheld the lower court's award of $215,000 as punitive damages. The court held that this amount was not excessive in light of the builder's and developer's total assets. Claims for punitive damages are part of tort claims to punish a defendant for outrageous conduct, as was ruled in this case.

C. DELAYS IN THE PROJECT

As stated in the previous section of this chapter, construction contracts will almost always have a completion date. The contractor may legally be assessed damage claims from the owner if he fails to turn over a completed facility to the owner on that date. The completion date may be extended under certain circumstances without penalty to the contractor.

The contract may say the project is to be completed in a given number of working days instead of a calendar date. If the contractor's forces cannot work because of weather or other action beyond his control, then these nonworking days, together with Saturday, Sunday, and holidays, are not part of the working days. Contracts that have a fixed completion date can be legally extended by any one of the following reasons.

1. *Unusually bad weather.* The contractor is not permitted to extend his completion date for "normal" weather he is supposed to have considered in his contract. Abnormal, so-called "Acts of God" would be tornadoes, hurricanes, floods, fires, mud-slides, etc. A rainfall or snowfall that would be considered by the weather bureau as possibly occurring only once in fifty years or longer would be considered justification for contract extension.

2. *A delay of work due to inability to receive materials and equipment.* This has caused questions and court action with regard to justification for extension of completion of a project. If failure to receive materials because of poor organization in not allowing sufficient time for delivery is the reason, then an extension is not justified. However, if delayed delivery is due to unforeseen strikes, or economic conditions that prohibit a normal delivery time period, then the contract period can be justifiably lengthened.

3. *A strike of work forces of the contractor or a subcontractor.* A contractor may have a weak legal position if the owner can prove that the strike was due to unethical, arbitrary, or capricious action on the part of the contractor or his personnel.

4. *The project is delayed by the owner.* The owner may fail to comply with his duties as described in the contract. Some of the duties may be to (a) prepare the site, (b) supply utilities, (c) supply materials or equipment, (d) provide police or fire protection, or (e) make payments on time.

5. *The owner may want to stop or delay the project* because of a number of reasons.

6. *Illegal or fraudulent action* on the part of any party connected with the project other than the contractor.

If there is a disagreement between the owner and contractor on a justifiable extension of completion date, the parties will have to seek arbitration or a decision from the court. An honest give-and-take agreement is the wisest decision in most cases.

Contracts should contain a "no-damage" clause which grants an extension of time for a delay in construction but denies any extra compensation. Such clauses, which have been upheld by the courts, prevent the contractor from making a monetary claim against the owner every time he is delayed. In an extreme case, where the contractor can unquestionably prove financial hardship, the courts have granted the contractor extra compensation even when the contract contains a "no-damage" clause.

D. TERMINATION

A contract is terminated when all parties have completed their responsibilities, when one or more of the parties have breached their part of the contract, or where it is impossible to continue performance. In the majority of contracts, termination takes place for the first of the above stated reasons. A contract remains in force until both parties have fulfilled their obligations or until one of the other reasons for termination takes place and a legal settlement is made. When one party to a contract fails to continue their obligations while the second party is able and willing to perform, then the contract is breached by the first party. Since the first party has breached, the second party is no longer obligated to continue their part of the obligation. Any substantial default in performance discharges the other party and allows him the legal right to consider the contract terminated. The remedy for the second party would be to sue the delinquent party for damages. In a construction contract the owner cannot force the builder through legal action to complete the project. If the builder cannot or will not do so then the owner can sue for damages or call for payment of the performance bond by the surety (*see* Chapter 10). If the owner defaults in payment, then the builder will proceed with legal action to place a lien against the property. With a lien on the property, it can be sold by the court and with the proceeds of the sale, the builder will be paid for his costs plus loss of profit and any other proven damages.

In rare cases, it becomes impossible for a builder to proceed with the project. A contract is considered dissolved when, due to circumstances beyond the control of either party to a contract, the subject matter no longer exists. This is rare in construction contracts, but could happen when the uncompleted facility is destroyed by fire or a natural happening such as earthquake, flood, tornado, etc. Insurance could possibly provide the funds to reclaim or begin the project anew. However, a new contract would be required.

An example of a termination due to conditions incapable of execution would be that a new building is to be constructed adjacent to an older building. The contract specifications gave the means of necessary protection to the foundation of the older adjacent building. Upon the beginning of excavation, it becomes apparent and is verified by experts, that the procedure specified cannot be followed without extensive damage to the old building or construction methods that are prohibitive in cost. Because the work cannot proceed as outlined, the contract is terminated. The owner would be obligated to pay the contractor for any expense.

A new contract mutually agreeable to all parties will terminate an existing contract. In fact, a contract can be terminated at any time if this is mutually agreeable to all parties.

Some contracts will have a right-of-termination clause. Such a clause will usually be written in favor of the owner. Most government contracts will have clauses giving the owner the right to stop work and terminate the contract. Such contract provisions are called "termination-for-convenience" clauses. The provision in the contract gives the owner permission to withdraw from the contract when he wishes to. A part of such written contracts is that the contractor will be paid for all his costs incurred plus his profit on the actually performed work. The engineer or architect would be required to evaluate the percentage of the project completed for determination of the percent profit due the contractor.

Authorization for the owner to terminate the employment of a contractor and to take over the completion of the work is contained in most construction contracts. The owner is usually given authority to take possession of equipment and tools. Taking possession of equipment and tools may pose a complex legal problem, since such items may not be owned by the contractor, but by lessors or even employees. However, for the owner to continue work under a new organization, it may be very disruptive to the project if equipment is not retained. This would be the case of asphalt plants, concrete mixing facilities, and similar large installations that would be expensive and time consuming to replace.

The drastic action of removing a general contractor or subcontractor from a project would be contingent upon the contractor's failure to perform. The procedure for this action should be detailed in the contract. The first step would be for the owner to receive written notice from the engineer/architect or construction manager who was overseeing the project that the contractor was not proceeding in an expeditious manner or the work was faulty or of poor quality. Another reason for removing a contractor from the project would be failure to pay bills, labor, or subcontractors—signs of bankruptcy. Failure of the contractor to conform to governmental laws or ordinances is also a cause for termination.

The reasons that give the owner power to terminate and remove a contractor from a project should be detailed in the construction contract. Courts have upheld the legal right of the owner to remove a contractor when such conditions are given in the contract documents and there is evidence that the contractor is non-conforming. It is proper and necessary that the contractor be given at least seven days' notice prior to termination. Good overseeing of the project by the owner or his agent would provide early detection if all is not going well with a project. When such signs first appear, the contractor should be called into conference and minutes of such a meeting taken and distributed to all concerned parties. Large projects will require weekly or maybe daily meetings of top management of the owner, engineer/architect, contractor, and possibly subcontractors.

A contractor or the owner may wish to terminate a construction contract when the situation is not developing the way he intended it to go when the contract was signed. One of the parties may claim that the contract is not legal. Such claims can be made on the basis of several reasons. Requirements for a valid contract are discussed in Chapter 7.

Although by far the largest majority of construction projects are successfully completed, some end up in litigation. The most common reasons for such legal action are claims of misrepresentation, fraud, or mistake. Misrepresentation is a form of fraud. If intent to deceive can be proven, the courts will most likely rule in favor of termination of the contract. Suing and collecting for damages is another story, and the success of such action will depend upon many factors, including comparative skill of legal counsel.

Proof of fraud must be well estblished, but fraud is not condoned in any form in the courts. Unfair methods used in obtaining any kind of a contract can lead to serious pitfalls for the party which obtained a contract under false practices or pretenses. The courts do not look upon so-called "shady practices" with any degree of forgiving. Public opinion has rightfully been changing over many years from indifference to unethical business practices to unwillingness to tolerate such activity.

The party declaring fraud on the part of the other party to a contract must give prompt notice of the discovery of such and declare his intent to terminate the contract. He must not fulfill any duties to the contract after he discovers the fraudulent action. There are legal technicalities to claims of fraud and experienced legal advice must be followed.

If fraud can be proven, not only can the contract be cancelled, but both compensatory and punitive damages can be collected from the defendant. Since a proven case of fraud can be very costly, all parties entering into a contract must be very careful that all information is honestly and correctly provided. Although an unintentional mistake may under certain circumstances be cause for voiding a contract, it is not grounds for a claim of fraud.

E. CHANGE ORDERS

Almost all construction projects will require changes in the work during the construction period. A formalized procedure should be instituted and covered in the contract. The written notice of change in the scope or detail of the work is called a *change order*. As stated in Chapter 9, any change in the work of a contract may require a change in the contract price. Minor changes, such as substitution of material of like cost, will require no modification in contract price.

Initial agreement between the owner and contractor on any price change should be made before the work is performed. If this is not possible, the owner is protected from unreasonable charges by a contractor by contract law. In case such change order charges are brought into court litigation, the contractor will be allowed to charge only what is fair and reasonable. In case of a dispute on correct charge, arbitration by an independent third party is preferable to court action.

There have been many court cases wherein the owner or his agent gave verbal order for scope of work changes and the contractor responded to these orders even though the contract contained a clause prohibiting any change in the project unless given in writing. The courts have not been consistent in their rulings on whether a contractor gets paid for work done under a verbal order when a written change order is required by the terms of the contract. There is a difference of rulings on this legal technicality between states. Some courts have ruled that if both owner and contractor orally agreed to the change, then it is a binding contract notwithstanding a requirement in the general contract that written orders are necessary to change the scope of the work.

Owners should realize that even though they are paying for the construction there is a formal correct procedure (in writing) that should be followed in making any work changes. A contractor should also reply to any verbal orders in work changes with a polite "give me a written change order." Such a procedure can prevent difficult and costly legal entanglements. Some contractors have found, to their sorrow, that they have been unable to collect for extra work because they performed the work without written authority.

Writers of contracts should be aware that courts have ruled that there is a distinction between the words "extra work" and "additional work." "Extra work" has been defined as work beyond and independent of the contract and an oral agreement is binding and not restricted or effected by any contract statements. "Additional work" is related to the contract and is subject to the wording of the contract in progress and is evidently needed for the completion of the contract.

When an order is given for additional work that is listed as a pay item in the unit price schedule, the unit price will prevail for any quantity additions or deletions of the work.

When any question arises about payment for extra work or additional work, ethical considerations should prevail. If the owner is getting more than he bargained for in the original contract and he desires the additional work, then he should pay for it despite any advantage he may gain from refusing to do so on legal, technical grounds. If the contractor made a change in materials or extent of the work because it was to his advantage and without any request from the owner, then the contractor should stand the cost even though the new materials may be of a higher quality. A contractor should be aware of any "nonapproved" changes. If not accepted by the owner or his agent, the contractor may be required to undo the work and replace the material.

The owner, unless specifically prohibited by the terms of the contract, can hire another contractor to perform the extra work. Therefore, the on-site contractor is restrained from charging unduly high prices for extra work.

F. CONDITIONS DIFFERENT FROM THOSE SHOWN ON PLANS OR IN SPECIFICATIONS

The author was brought up on "horror stories" telling of honest, hardworking contractors going broke because of a highway cut requiring heavy rock removal when the contractor had guessed otherwise in preparing his bid. The uncertainties of what was below the surface of the ground were faced by contractors up to the early decades of the 20th century, but the development of subsurface exploration equipment has removed most of the guesswork from excavation activities. The design engineer on a project—whether road, bridge, building, etc.—would require an understanding of the subsurface conditions in order to prepare his design. This information should always be made available to the contractor. If the bidders on a project do not have all the information they think is necessary for them to calculate costs, they will have to prepare a high bid in order to provide financial protection. There is always some element of chance in bidding a construction contract, but the owner and engineer should provide as much data as practical. There may be occasions when the owner will try to judge the cost of gathering more data against the higher bids he may receive when that data is lacking.

The owner and his agent should never hold back from the prospective bidders any information that they have about the project. If the contractor later determines that this action took place, he has solid grounds for a suit asking for additional payment if he can prove that this lack of information increased his cost of construction, or that he would have entered a higher bid if he had had this additional knowledge.

A log of the test borings should be included in the design drawings. The log should show location of test holes, types of soil and elevation where the soil type changes, number of blows at various elevations to drive the soil sampler, and the location of ground water. The dates of the borings should be shown. This date has a relation to the ground water elevation.

A court case resulted from the fact that the ground water level was not shown on the test boring log. The contractor assumed that ground water was below the depth of boring or at least below the required excavation. When this did not prove to be the case, the contractor asked for additional renumeration and was refused by the owner. In subsequent court action, the court ruling was in favor of the contractor, since the owner withheld information that had a definite bearing on the bid price. The drill cores from the subsurface exploration should be retained at some location for inspection by prospective bidders.

A contractor sued a state highway department, claiming an extra cost of $724,000 to cover the additional cost of removing rock. The plans and specification estimated a certain quantity of rock excavation for storm sewers. More rock was encountered than estimated, and the contractor's on-site inspection did not indicate the quantity of rock encountered. The highway department had made some test boring in the area and a log of such was available to the contractor before bidding; however, he made no request for the report. The court denied the contractor's claim, ruling that there was no warranty that the plans disclosed all the information known to the department.

The above case indicates that the highway department did withhold information from bidders which it had, and possible disclosure may have affected the bidding. One wonders about the justice of the court ruling in this case. The mitigating circumstance may have been the fact that the contractor did not seek all the information that was available to him. The degree to which the highway department made known that it had taken test borings is not known.

A claim for extra money is made by a contractor based on changed conditions. A factor having a major bearing on whether this claim will be honored by the engineer/architect is whether the contractor should have really anticipated the conditions. When claims indicate that the contractor showed lack of knowledge or experience, the claim will most likely be rejected by not only the owner, but also the court. Changes in conditions are almost inevitable; some should be anticipated by a contractor. Those that should be anticipated usually result in a rejection of claims for extra money.

The specifications usually state that a bidder will visit the site and become familiar with all conditions. This responsibility is placed upon a bidder. By this action, the contractor is prevented from making a claim against the owner for site conditions of which he was not aware.

There may be other conditions that are different from what is shown on the plans. If they are due to an error on the part of the design professional, then the contractor may have a legitimate reason for asking for an increase in his bid price. He must, however, prove that the change from the drawn or specified conditions to the actual conditions will cause him an increase in cost to construct the facility. The owner will rely on the design professional to verify any claim from the contractor. The engineer/architect should be very fair and honest in evaluating any claims for extra compensation. In no way should he be unduly influenced by his feelings toward the contractor or owner. Strictly ethical action is required. If the contractor does not agree with the design professional or owner on their evaluation of the claim for additional compensation, then he has recourse to mediation, arbitration, or a court settlement. If the claim for extra compensation is small, then a claim through the courts is usually not worth the time and cost. Simple arbitration may be the best route.

G. PROTECTION OF WORK

Until the facility is completed and accepted by the owner, the project, including all materials on site, is the responsibility of the contractor. Even though the contractor has received progress payments for construction completed or for materials at the site, the facility is not yet legally the property of the owner. The contractor should and must provide all feasible and necessary protection from storm, fire, theft, vandalism, etc.

A contract should contain a "protection of the work" clause to the effect that all work and materials on the site should be protected from financial loss by builder's risk. All subcontractor contracts should require the same. The owner or his agent should require that such insurance policies be retained by the owner as proof that there is such insurance and the premiums have been paid.

The owner could take the attitude that the safekeeping of the project was the responsibility of the contractor and that in case of loss the contractor has to replace the loss. However, in case of a disaster that destroys a large part or all of the completed facility, the contractor is not likely to have sufficient capital for replacement and may declare bankruptcy. Insurance protects the owner from this loss which he would have if the contractor could not replace the damage with his own resources.

A question arises as to the obligation of the contractor to begin anew and complete a project when total or heavy destruction has taken place. If a partially completed structure is totally destroyed by fire or acts of nature, court rules have not relieved the contractor of his contract to build the facility. As long as the destruction does not make it impossible to build the structure, then most likely the contractor must begin again. Extension of time and compensation to remove the debris is reasonable. The structure should be covered by insurance.

If the construction had involved a considerable amount of time, starting over may be a financial hardship to the contractor. Prices may have escalated in the previous construction period. The contractor may not wish to be involved again and also the owner may no longer want the facility, or a considerable change may be desired. A new, mutually agreeable contract can be written replacing the old one. The former contract is then terminated.

If the contractor wanted out of a contract, in which the constructed portion had been destroyed, and the owner required it to be rebuilt, the fair price for rebuilding would enter into any court litigation. It is not likely that the courts would require the contractor to take a large loss in rebuilding if the prices had escalated since the first contract was signed. A new negotiated price for rebuilding would be the reasonable approach.

The owner or his agent, when inspecting completed work or determining progress, should be observant for hazardous conditions such as would cause fire, water damage, or injury to persons. Even though the owner is covered against loss by insurance, any calamity would result in a considerable time delay in the progress of the project. Any hazardous conditions should be reported to the contractor's superintendent and corrections noted. The contract should contain a clause giving the owner this power of requiring the contractor to correct any hazardous conditions, but not relieving the contractor of the responsibility of detecting and correcting hazardous conditions.

H. SUBSTITUTION OF MATERIALS

The drawings and specifications should clearly define the quality of all materials to be used in a constructed facility. However, after the contract is signed it may become necessary to substitute a different quality of material from that specified. This could be due to a number of reasons, such as short supply; new, improved materials on the market; change of mind of owner; derogatory information on experience of material specified; etc.

The contractor may initiate the request for a substitution, especially when there is a short supply and a wait on delivery may slow the progress, or in extreme cases stop the work. It may be possible that the price of the material specified has increased greatly and a different type or brand that has not increased in price will do as well. Of course, all changes must be approved by the engineer/architect. The engineer/architect should be reasonable in approving a substitution. In the case of short supply, finding a substitute material will be necessary. With a considerable increase in price, the engineer/architect will want to use good judgment and not just release the contractor because of a quirk in the market that was not anticipated. If the contractor made a poor estimate of cost in his bid, it is no reason to allow a substitution unless the requested substitute material is equal or superior in quality.

If the owner or his agent initiates the request for a substitution, then the contractor must make the change. Such should be stated in the contract. Any change in cost either up or down should be added or subtracted from the total contract price. Changes should be made as early in the project as possible to avoid any undue cost to do rework, return materials, change the equipment, or delay the project.

I. LIENS

The general subject of liens will be discussed in this section. It is too complex a subject to cover in detail within the scope of this book. Gifis' *Law Dictionary* [1] defines a *lien* as follows: "a charge, hold, or claim upon the property of another as security for some debt or charge. . . , not a title to property, but rather a charge upon it; the term connotes the right which the law gives to have a debt satisfied out of the property."

A lien is a legal writ that a person obtains against a property as a result of a debt owed to a person by the owner of the property. The debt must come as a result of prior ownership of the property by the claimant; unpaid bills for materials used in or on the property, or unpaid wages for doing work on the property. An important aspect of a lien is that the property may or may not be in possession of the person to whom the debt is due. Also of note is that a lien can remain against a piece of property even though it is sold by the debtor. A person may or may not be reluctant to buy a piece of property which has a lien against it.

There are many classifications of liens. One broad classification is private and public liens. The distinction here is whether the charge is against private or public projects. A well known lien against private property is that of the automobile purchased with loan money. The lender will attach a lien to the car by a recording of such on the title. If the owner sells the car, then he must pay the lien holder first before the money can be diverted to any other use. Failure to do so is a crime.

There are many other types of liens: tax, mechanics, materialman, judgment, and attachment. All of these types of liens are subject to special laws and such laws vary from state to state. More details can be found in other publications.

The two types of liens that most commonly apply to construction projects are the mechanic's lien and the materialman lien. Under a mechanic's lien, a workman of any grade can obtain a lien against a property on which he worked if he has not been paid by the contractor. Since the lien is against the owner's property, the owner would have to pay any delinquent wages to clear the lien if the contractor does not pay his workmen. An owner on a large project protects his property against a lien by: (1) Not paying the contractor unless he certifies that all bills, including labor, have been paid, (2) Requiring the contractor to purchase a payment bond (*see* Chapter 10). A homeowner or an owner of a small project may wish protection by procedure 1 above.

Mechanic's liens take precedence over mortgages, deeds of trust, or any other types of encumbrances.

The materialman lien is a legal charge against the property because of default in payment of costs of materials used in the construction or repair of the property. The material must have been used in the specific property being requested for a lien.

There are certain statutes that govern the filing of a lien. A lien is usually processed by filing with the county recorder of the county, where the property is located, a written claim that contains the necessary detailed information as required by the laws of the state. There is usually a time limit on filing a lien after ceasing to render services or the furnishing of materials.

After the lien has been filed there is a time limit (usually not longer than one year) wherein an action to enforce must begin. The action to enforce usually requires an appearance before a court wherein all the evidence of work done or materials furnished is presented. If the judge rules the claims are valid, then an order is given by the judge to the sheriff to sell the property at a sheriff's sale and use the proceeds of the sale to pay off all lien holders and all court and legal costs. The owner has a time limit, as prescribed by statute, after the court order in which he can void the liens and the sheriff's sale order from the court. To do so at this time period, he must pay off the amounts of the liens plus any court costs and legal fees encumbered by the lien holders.

When a property is sold at a sheriff's sale, it is by auction. The property goes to the highest bidder. Usually the owner himself can bid on the property and buy it back if he is the highest bidder. The proceeds of the sheriff's sale go first to pay for the court costs (including sheriff's sale) and then the remainder pays the authorized claims of the lien holders. If the remaining proceeds of the sale are not sufficient to cover all lien claims, then the mechanic's liens are paid first and what is left is prorated among the materialman's liens.

After the sheriff's sale and the distribution of the proceeds of the sale, even though a lien holder may only get ten cents of each dollar in his claim, there are no further claims against the property. The high bidder gets clear title to the property for whatever price he bid.

If the proceeds of the sheriff's sale are greater than the court costs plus the total lien claims, then the balance is paid to the owner in default.

The total process of lien claims and court action usually requires legal counsel. State laws and statute of limitations vary from state to state. Failure to take the proper prescribed legal action can cause a loss of lien.

An interesting case of lien action took place in Utah several years ago. A developer began the construction of a recreation facility and sold memberships. Part of the facility was built and used by the members. The developer acted as contractor and did the work by hiring subcontractors. After a period of time, it became evident that the developer had used only a limited percentage of the membership fees to pay the subcontractors for the construction and was considerably in debt. The subcontractors then entered liens against the property. In the heat of confrontation by the members, the developer left town and upon legal advice, the members were told they had no equity in the property since their agreement with the developer was only a use permit and not any right of ownership. The members then appealed to the lien holders to hold off on court action and accept seventy-five cents on the dollar, as well as to allow time for the members to raise these necessary funds. Through the advice of the lien holders' legal counsel, they refused the appeal from the members and the property went to sheriff's sale. The members raised some money for bidding at the sale, but had limited funds because of the time limitation as well as bad publicity.

The result of the sheriff's sale was that the lien holders received less than fifty cents on the dollar, out of which they had to pay attorney fees. The members lost what they had paid in membership fees and only gained a legal education. The hasty action (supported by their attorneys) cost the lien holders money. The only ones to gain from the total action were the person who was high bidder at the sheriff's sale and the lawyers involved. The developer may have gained, but since the county attorney took no action, an accountability of the membership fees was never realized.

In some states a design professional may attach a lien against a property he designed if he was not paid all of his fee. In other states the design professional does not have this right. In such states, the reason behind such a negative judgment is that the designer had supplied no direct labor or material to the construction.

J. FREE ENGINEERING OR ARCHITECTURAL SERVICES

A practice that has been somewhat common in the construction industry is the giving of free engineering services by suppliers of equipment. This type of conduct has been common among house contractors wherein a contractor supplies free plans if he is given the contract to build without competitive bidding and with either a lump-sum bid or a cost-plus contract. The owner really does not know what amount he is paying for the plans since there is no competitive bidding. Such plans are most often a "manufactured" set and not unique to the desires and wishes of the homeowner.

In moderate size construction projects, the design professional may be approached by equipment suppliers who will furnish all necessary drawings for the installation of the type of equipment they sell. Air conditioning is a typical example. Such an offer may be very tempting to the engineer/architect, since it can save him considerable design time. When such "free" plans are prepared, they will almost always specify the one type of equipment which the provider of the free plans sells. Even if the specifications contain an "or equal" clause, the plans may be prepared in such a manner that only one kind of equipment meets all the details of the plans and specifications.

In public projects, such design procedures may be entirely unlawful. For private projects, the use of "free" designs may be unethical. The design professional should consult with the owner if he believes the use of such designs would be to the advantage of the project. It will usually be to the advantage of the project only if there still remains free competition in the bidding for the supply of the subject equipment. As the old saying goes, "There is nothing free in the world." Someone pays for everything and the design professional should be accountable for who pays for what and how much.

K. SUPPLY CONTRACTS

Construction contractors and subcontractors are always very involved in supply contracts. The design professional is only involved in specifying the quality and quantity of material and approving the material the contractor installs in the project.

The proper estimating of material costs, judicious purchasing, and correct timing of delivery can make the difference between a profitable or a losing contract. The contractor will want to submit purchase orders on a fixed-price basis early in the contract period. However, he will want to have delivery only as he needs the materials. Assurance of delivery at the required time is of great importance, but is not always possible. At most construction sites, there is usually limited space to store materials, especially space where the material will be protected from the weather, theft, or other damage hazards. In a rising price market, the contractor may have to take early delivery and obtain warehouse space for storage of materials until needed. This double handling will add to the costs of the contractor.

As the contractor would like to fix the price of all materials as early in the project as possible, he will usually want as many fixed-price contracts as possible. This will be especially so during an inflation period such as the construction industry has experienced for several decades. A fixed-price contract binds the supplier to supply the material at a fixed price on a given date. The delivery date may have a few days' flexibility. The supplier agrees to the fixed price regardless of what happens to the market.

Most states have a law for supply contracts. This standard law is the Uniform Commercial Code. This Code requires that for all orders over $500, the contract must be in writing. Phone orders are contracts if a price and delivery date has been established. However, to be easily enforceable there should be a followup letter or price quotation. The courts have held suppliers to oral contracts where partial deliveries have been made, and partial payments at the orally agreed-upon price have been paid. Partial performance of a contract may be sufficient in the eyes of the court to show the existence of a contract.

Fixed-price contracts are binding on a contractor as well as a supplier. If the market price has dropped below the contract price by delivery time, the contractor must nevertheless pay the higher contract price. An expert purchasing agent who follows the market trends and makes correct decisions as to types and timing of purchases is a very valuable employee for a contractor.

If a general contract between owner and contractor contains required warranties for equipment installed, the contractor should obtain the same warranties in writing from the suppliers of said equipment. The extent of the warranties should be sufficient to cover the contractor. Both the owner and the contractor should receive and retain copies of all warranties.

An important aspect of purchase-order contracts is to clearly define the location of delivery and who pays the shipping costs. The purchase order and price quotation, including the supply contract, will give the F.O.B. location. The F.O.B. location is usually the manufacturer's or the supplier's warehouse. When this is the case, the contractor will pay all shipping costs. If the contractor is to pick up the material with his own trucks, F.O.B. the plant site is preferable. If the contractor wishes a delivered price, then he will ask for a quotation based on F.O.B. the construction site.

As previously discussed, the payment schedule of the owner to the contractor is an important part of any construction contract. The owner may find it necessary to help finance the project by paying for materials upon delivery yet well before installation. This payment procedure may yield lower overall construction bids. It may also be necessary to pay for equipment while it is being manufactured. This will be the case when such equipment is costly and has a long time period of manufacture or fabrication. Structural steel for a large building or bridge is an example. The owner may save money by financing the fabrication of the steel rather than requiring the general contractor to do so.

There are dangers in paying in advance for material before it is actually incorporated into a project. If the contractor is in financial trouble, there may be claims against the materials by the contractor's creditors. If the contractor goes into bankruptcy there may be legal claims against the material by the contractor's creditors if it has not been incorporated into the facility. This could be the case especially if the material is stored off the site. The owner's payment to the contractor for the material may not necessarily give the owner legal title to it.

Loss of any material through fire, theft, etc., when it has been paid for by the owner, can result in the owner's loss and not that of the contractor. Insurance against loss should be the responsibility of the owner when he pays for the material before it is part of the completed construction.

Because of possible legal complications, the owner should generally not pay for any material or equipment before it is installed in the facility.

L. THE AUTHORITY OF THE ENGINEER/ARCHITECT

The design professional usually has the role of not only preparing the plans and specifications but also the responsibility of approving and certifying the work of the contractor. He may also have the role of judge on questions of duties and responsibility of the contractor. The contract between the owner and the contractor should clearly define the role of the engineer/architect in administration of the contract between owner and contractor. If the engineer/architect has the power to decide interpretations of the construction contract, it should be so stated. The engineer/architect may be considered a neutral figure in the interchange between owner and contractor. He should try and play a neutral role between the two principles to the construction contract. This neutral role may be somewhat difficult, since the engineer/architect is in reality an agent of the owner and can be discharged by the owner, but not by the contractor.

Actions of the engineer/architect in deciding interpretation questions and in the judging of performance have been considered by the courts as binding upon the owner. The contract should specify such authority. Responsibility without authority is an onerous position to be in, and the engineer/architect should avoid such a situation.

During the construction of a project, there is certain information that may need transmittal to the owner. The contractor may transmit such information to the engineer/architect. There have been cases where such was done and the owner denied receipt of such because the contract did not specify that notices be given to the engineer/architect. Courts have held that the engineer/architect has authority to receive such notices from the contractor, even though this is not spelled out in the contract. The engineer/architect is considered an *implied* agent in such cases. The efficient engineer/architect should be proficient in transmitting to the owner any communication received from the contractor.

Courts rarely apply the concept of implied agency in the case of contracts between a contractor and a public agency. Government contracts are often more precise in delegating authority. Such contracts usually specify only one person who has the authority to act as the agent of the goverment. This person or office is designated as the contracting officer. All communications pertaining to the contract must be transmitted to and from the contracting officer. This is the case even though the government agency hires an engineering firm to be the design professional and to monitor the construction.

M. MEDIATION AND ARBITRATION

Mediation and arbitration of disputes between parties has been going on in society ever since man first began to live together in communities. Such mediation or arbitration over the centuries has been more or less informal, with a friend, governmental authority, religious leader, etc., acting as the mediator in disputes. Of course, our system of courts developed from this pattern of mediation.

In recent years in America, a formal procedure has developed for mediation and arbitration of disputes. This arbitration procedure is best known by the public in labor disputes. However, it is a very accepted practice in the construction industry. The procedure has become formalized and many construction contracts now have specific stated provisions in the contract to refer disputes to mediation, and/or abitration.

Prior to 1966, construction industry arbitration rules were under the American Institute of Architects Conditions in an informal manner or under the American Arbitration Association's (AAA) Commercial Rules. The informality of procedures and lack of rules resulted in some dissatisfaction. After a comprehensive study by engineers and architects, it was concluded that the procedure could be improved. A new set of rules for arbitration of construction contracts was drawn up and adopted in 1966. These rules are endorsed by ten separate organizations, including the American Society of Civil Engineers, American Institute of Architects, National Society of Professional Engineers, and Associated General Contractors. The rules and procedures are administered by the AAA.

The purpose of mediation and arbitration clauses in construction contracts is to arrive at an equitable solution to contract disputes without resorting to costly and lengthy court settlements. It has many advantages over court litigation. The primary advantages are:

1. disputes can be handled without costly legal counsel and court action;
2. judgment will be by experts in the field of engineering, architecture, and construction;
3. disputes can be settled quickly and finally.

Mediation consists of a procedure whereby the effort of an individual or several individuals assist the disputing parties in reaching an equitable settlement of a dispute without any formal procedures. The AAA has a framework to administer the mediation process in an orderly, economical, and expeditious manner. They will identify individuals who are qualified and experienced in the mediation process and who are knowledgeable in the field of construction.

The mediation process can easily be put into action if the contract contains a mediation clause. A clause suggested by the AAA reads as follows:

> If a dispute arises out of or relating to this contract or the breach thereof, and if said dispute cannot be settled through direct discussions, the parties may agree to first endeavor to settle the dispute in an amicable manner by mediation under the Voluntary Construction Mediation Rules of the American Arbitration Association, before having recourse to arbitration or a judicial forum.

After a dispute has arisen, the parties to the dispute can then enter into and sign an agreement which, as suggested by the AAA, would read as follows:

> The parties hereby submit the following dispute to mediation under the Voluntary Construction Mediation Rules of the American Arbitration Association. The requirement of filing a notice of claim with respect to the dispute submitted to mediation shall be suspended until the conclusion of the mediation process.

The clause can also provide for the number of mediators, their compensation, method of payment, locale of meetings and any other item of concern to the parties.

The detail procedures and rules for mediation can be obtained from the AAA. Only the general procedure is given here. Parties that have or have not signed the mediation agreement in their contract can initiate mediation by filing a written agreement to mediate a dispute. Details of the dispute are in the request for mediation. The AAA then appoints one or more mediators from a qualified list. The parties to a contract can reject any listed mediator. A mediator who has any personal interest in the result of the dispute must disqualify himself as a mediator. Strong ethical considerations demand such action.

The mediator or mediators work independently of AAA once they have been chosen. The mediators are paid a reasonable fee jointly by the parties in the dispute. Expert witnesses can be called in, providing the parties agree and share the expense. If a party produces a witness, the expense is carried by that party. Mediation procedures are private affairs void of publicity.

The decision of the mediator is not legally binding on any party. The process is a means of amicable settlement of contract disputes. It is a satisfactory way of settling minor and sometimes major contract disputes.

Arbitration is a more formal and involved procedure for settling disputes. The AAA has, through the help of the sponsoring organizations, developed Construction Industry Arbitration Rules. The latest rules were effective April, 1979, and can be obtained from the AAA in New York City.

Arbitration can be provided in the original contract by the following *future dispute arbitration clause:*

> Any controversy or claim arising out of or relating to this contract, or the breach thereof, shall be settled by arbitration in accordance with the Construction Industry Arbitration Rules of the American Arbitration Association, and judgment upon the award rendered by the Arbitrator(s) may be entered in any Court having jurisdiction thereof.

Arbitration is regulated in many states by state law. Their legal procedures are carefully followed under the rules of the AAA. If the rules are followed and the contract has an arbitration clause, the finding and award as determined by the arbitrators can be legally binding. At one time the courts were hostile to arbitration and some states did not permit it. However, this has changed and the arbitrator's decision has been given substantial finality.

Arbitration does replace court litigation and has some decided advantages and disadvantages. Sweet [2] gives details with pros and cons to this way of deciding disputes.

Arbitrators can be selected through a panel provided by an organization or by the individual parties to the dispute. Arbitrators will act similar to judges in a court trial. They will not, however, be as restricted in procedure as court judges. Usually the legal rules of evidence need not apply. Anyone involved in the procedure can question witnesses. Either parties may have legal counsel present but the usual courtroom maneuver of attorneys is absent. Testimony may or may not be recorded. In complex cases involving large monetary claims, testimony will most likely be recorded.

Under the AAA rules arbitrators are paid only for expenses the first two days, but if the hearing goes beyond two days then they must be paid by the parties to the dispute. Arbitrators are not strictly held to the contract but can base their decisions on the broad standard of equity.

A considerable advantage of arbitration over court litigation is the length of time to settle a dispute. Unless the dispute involves testimony from many witnesses whose appearance at a hearing has to be scheduled, an arbitrated case can be processed much more quickly than a court trial. Trials have to be fitted into loaded court schedules and can drag on for a considerable period of time.

Arbitration decisions can be appealed to the courts and some past cases have been. Court rulings will uphold the findings of the arbitration unless:

1. there was fraud of any kind in the arbitration process;
2. the decision was outside the limit of issues submitted for arbitration;
3. all parties did not have a fair and full hearing and were restricted from presenting evidence.

If disputing parties have entered into an arbitration proceeding voluntarily and have agreed to submit the issues to arbitration, then the judgment should be accepted as final unless one of the three above listed actions took place.

The May-June 1978 issue of *News and Views*, published by AAA, contained a questionnaire to be answered by the readership. Nine hundred and four questionnaries were returned. Thirty percent identified themselves as experienced in construction arbitration procedures. The results of the survey were published in the Spring 1979 issue of AAA's publication *Punch List*. The results of this survey as published were as follows:

	% Answering Yes
Arbitration is speedier than litigation	94
Arbitration costs less than litigation	92
Arbitrators tend to know more about practical matters than judges	63

Arbitrators tend to be more knowledgeable than juries	89
An arbitration hearing is less threatening to parties and witnesses than a trial in court	39
Arbitrators are more likely to compromise than a jury	48
Arbitrators tend to ignore the law applicable to the case	12
An arbitrator should attempt to mediate	35
It is advisable to have a lawyer representing you in arbitration	70
Lawyers will generally advise their clients not to provide for arbitration in their contracts	19
Business people tend to prefer arbitrating to going to court	60

Arbitration is gaining acceptance in construction contracts and will very likely become a more accepted route in the future than court litigation. In most cases the advantages of speedy resolving of disputes, less cost, and decisions by knowledgeable and experienced arbitrators far outweigh any disadvantages in the arbitration process.

REFERENCES

1. S. H. Gifis, *Law Dictionary*. Barne's Educational Series, 1975.
2. J. Sweet, *Legal Aspects of Architecture, Engineering and the Construction Process*. St. Paul, Minn.: West Publishing Co., 1970.

12. MISCELLANEOUS ACTIVITIES OF ENGINEERS AND ARCHITECTS

A. PROPERTY

Almost all engineering construction contracts start with a given piece of property. In many instances the property is purchased for the particular purpose of constructing a facility on the land. An engineer may be very involved in acquiring a necessary piece of land. He may be asked to evaluate several possible sites for the placement of the facility.

In order to evaluate several sites and make recommendations to the owner, the engineer may have to make a number of different studies. Some of the most likely studies will include the following.

1. Reading of records at the county recorder to determine ownership of the land and extent of any encumbrances, the type of zoning, and whether a zoning change may be needed. The feasibility of obtaining a zone change should be evaluated. Some inquiry with property owners to determine possibility of purchase may be necessary. Time should not be wasted on unavailable property.
2. A survey to establish boundary lines of the property. A profile map may be necessary to study the possible layout of the facility. Ground slopes and elevations may be a major factor in the constructed cost.
3. A preliminary soil investigation with a study of the surface geology. A study of soil borings previously taken in adjoining property may give some information with regard to what might be underground. What is below ground surface is also a major factor in the cost of a construction project.
4. A study of utilities. This study would include the availability of water, electricity, natural gas, and sewer lines.

5. The study of forms of transportation into and out of the property. The adequacy of roads, streets, and railroads plus work involved in development of transportation within the property.
6. Investigation of possible environmental impact and local opposition or support to the facility being placed on a particular piece of land.

All of the investigations should be very thorough to avoid any legal complications that might develop. Purchase of land should be from owners who have clear title. Sometimes a clear title can easily be obtained by paying back taxes, lien claims, etc. The engineer may be able to handle all details of land acquisition for the owner or he may seek some legal help. A title search can be determined to ascertain the abstract of title which contains the recorded facts concerning the particular property. This title search is usually performed by an abstract company. A lawyer can then use the abstract to determine if the owner has marketable title. If a title is clear without any encumbrances, title insurance on the property can usually be purchased. This insurance will require the insurance company to defend the new owner against any unfounded claims on the property. The insurance will also cover any loss the owner may have due to legal claims against the property.

The laying of claims against real property (land and its attachments) has been discussed in the section on liens in Chapter 11. Mortgages and liens are means of attaching debts to property to force the payment of debts incurred by the property owner. Generally, only debts that were incurred by spending money on the property in the purchase, or in building a facility on the land can be claims against the land.

B. ZONING

Zoning of land is a common practice in the United States. Zoning is done by city or county governments and not by state or the federal government. The legal act of zoning land for specific use was established several decades ago, and zoning ordinances have been upheld by the U.S. Supreme Court.

Zoning ordinances divide the land into several use zones. These zones range through several grades of residential zones (single-family homes to multiple-residence apartment buildings) to commercial, manufacturing, and agricultural. Any facility built on the land must conform to the building requirements of that zone. For instance, a retail store could not be built in a residential zone area and a manufacturing plant could not be built in a residential or commercial zone and probably not in an agricultural zone. The purpose of zoning is to maintain an environment in keeping with the majority wishes of the landowners.

Zoning laws vary somewhat in detail from city to city. In addition to the prescribed use of the land, there are also regulations with respect to the clearance from property line to building, maximum building area with respect to area of lot, and setback from street, as well as other use regulations.

It is sometimes possible to obtain a permit to build even though the building may not meet all requirements of the zone. In order to do this, the owner will have to fill out the necessary forms and prepare the required drawing and maps. He will most likely have to obtain written approval from neighboring property owners. He will then have to appear before the zoning board together with all necessary documents. It may be much harder to obtain approval of a request for variance for change in use than for, say, a permit to build a carport closer to a property line than allowed by regulation. If the zoning variance request is for a major nonconforming use, the zoning board will hold a public hearing to determine the wishes of the landowners in the neighborhood. A request for nonconforming use will most likely meet opposition at this period of time. Citizens are more strong in their feelings than ever before in the history of the United States with concern for the environment, especially in their area of residence.

If a new area is zoned for the first time or an area is rezoned, then the property owners cannot be forced to change what they were previously using the land for. This use will then be classified as nonconforming use. For example, an area just outside the city limits is annexed into the city upon request of a majority of the landowners in the area. It is then zoned residential. However, some of the residents in the area keep animals such as chickens and horses, which constitutes nonconforming use for the residential zone. They cannot be required to give up the keeping of animals but can continue to live as before. If the property is sold, however, the new owners must conform to the zoning regulations. Also, if the residents give up their nonconforming use, they cannot then go back to the keeping of animals at some later date.

When anyone prepares to purchase land, they should thoroughly investigate the zoning requirements pertaining to that piece of property to make sure that their intended use is in conformity to the ordinances.

C. MORTGAGE

In acquiring real property, whether an individual for a home or a corporation for commercial or manufacturing building, it will usually be necessary to borrow money to build the facility and/or buy the land. In borrowing money for real property, there will be a written pledge to repay the borrowed money in accordance with some payment schedule. This written pledge in which the property is secured for repayment of the loan is called a mortgage. A mortgage will be for repayment of a given amount of money plus interest at a set rate and over a given period of time. Most mortgages have repayment at monthly intervals. Once a mortgage is signed, the interest rate or the time period cannot be changed except upon mutual agreement of both the lender (mortgagee) or the property owner (mortgagor).

In theory there are two types of mortgages. One is the "title" type of mortgage, wherein the mortgagee has acutal title to the property until the loan is repaid. Of course, the mortgagee cannot sell the property or transfer title unless the mortgagor defaults on payments.

The second type of mortgage is the "lien" type, wherein the lender has a lien on the property. If the mortgagor defaults on payments, the mortgagee can foreclose on the property and sell it for the amount due. Foreclosing and selling property has to conform to the laws of the state and must have judicial approval.

Which type of mortgage is used depends upon the state law. The "lien" type is the most common.

If a foreclosure takes place and the property is sold, then any difference between the selling price and the debt plus costs of selling goes to the owner. If a piece of property has a remaining debt of $35,000 and the property is foreclosed and sold for $50,000, the owner would get $15,000 less the legal costs of foreclosing and realtor's fees for selling. These additional costs will amount to several thousand dollars. If the debt is small, the lien holder may not waste much time in selling the property in order to close the action. If this is the case, the mortgagee may not seek the highest possible sale price. The borrower may lose money in such a case. The mortgagee cannot just "give away" the property however. If he sells at a price well below the known market price, the mortgagor could enter suit against the mortgagee claiming damages. If a borrower cannot meet payments, it may be to his financial advantage to find a buyer himself in order to obtain the highest possible price.

When a loan is taken with real property as security, there are two documents signed by the lender and borrower. One is the mortgage document which has to be recorded, for a small fee, with the county recorder. Anyone planning on purchasing real property can then determine by a search of the records at the county recorder's office if the property has a mortgage or any other type of lien against it.

The second document signed by the lender is a note of loan in which the borrower agrees to repay a loan of a given amount plus interest by a given date. This second document protects the lender in case the borrower defaults and the sale of the property does not cover the debt. The borrower is then still in debt to the lender for the difference. There has been a misconception by some buyers of homes that if they cannot make the payment, they can "walk away" and leave the property to the lender, and if their down payment was small and the monthly payments were less than rent then they have suffered no loss. In the case of a small down payment and falling prices, the lender may suffer a loss in case of default. However, any loss is still owed by the borrower and legal steps may be taken in an attempt to collect the remaining debt.

A mortgage and the accompanying debt may be sold by the mortgagee to a third person without consent of the mortgagor. However, the new mortgagee cannot change any terms of the mortgage or debt document.

Mortgages are great protection to borrowers in times of rapid inflation. This has been the experience of home owners who have purchased mortgages when interest rates were 5 to 7 percent and then in a matter of years rose to 8-10 percent. In such cases, lending institutions have offered discounts to borrowers if they would pay off their loans early. Many mortgage loans will have clauses that say there will be a penalty charge of a specific percentage of the unpaid balance if the loan is paid off early. This clause is a means of discouraging the borrower from seeking another lender at a lower interest rate, especially when interest rates are declining.

Another type of mortgage is called a *second mortgage*. The second mortgage is just as the name implies, a second mortgage on the land in addition to a first mortgage which is still active. A borrower may find that he can only borrow a certain amount of money on a first mortgage basis, but it is not enough for the improvements he plans. He then seeks additional money from a second lender who would take a second mortgage for the second loan on the property. A second mortgage will usually demand a higher interest rate because there is more risk in a second mortgage. A first mortgage has precedence over a second mortgage in case of foreclosure. On a foreclosure sale for nonpayment of a debt, the holder of the first mortgage would receive his unpaid debt plus costs before the second mortgagee would receive any money from the proceeds of the sale. If the proceeds of the sale were not sufficient to cover both debts plus legal fees and selling costs, the second mortgagee would suffer the loss.

A second mortgage route can sometimes be an advantage when a property owner wishes to sell a piece of real property in which he has a mortgage of an amount that is a small percentage of the value of the property. The owner finds a buyer who can assume the first mortgage but does not have sufficient funds to pay the owner the difference between the selling price and the amount due on the first mortgage. The owner then makes the sale possible by accepting a down payment and taking a second mortgage on the balance. In this way, he makes the financing possible for the buyer. His risk of losing his interest in the property by a default on the part of the new buyer is slight. If the new owner defaults his payments to both the first and second mortgagee, then the second mortgagee can pick up the payments to the first mortgagee to prevent the first mortgagee from foreclosing. The second mortgagee can foreclose on the new owner if he believes it is in his best interest to do so.

A person taking a second mortgage would want to be sure that the value of the property is worth more than the total of the first and second mortgage and the mortgagor is a good credit risk.

D. USE RIGHTS OF LAND

There are a number of kinds of rights a landowner can grant to others without losing the right or title to his land. These, called servitudes, will be discussed in order.

> 1. *Easement.* An easement is the right to use another person's land in a particular way. This is usually granted by a written document and for a consideration. The easement right should be contained in the deed to the property and shown on the plot plan in the Recorder's Office.

The most common of easements are those for the placement of public utility lines such as electric lines, power poles, water lines, etc. Most municipal governments will require the granting of easements along the back or front property lines of all lots in a new subdivision. For instance, if a utility easement was along the back five feet of a building lot, the owner retains title to this property; but a utility company could place poles or underground lines in this five feet wherever necessary. If it meant digging up some plants to do this, they would have that right. For good public relations they would do well to replace the plants, but would not be required to.

The right of easement usually goes with the land irrespective of the owner of the land. A special type of easement that does not transfer with title to the land is an easement in gross. An easement in gross is a right of use in land granted for the personal use of an individual. Allowing a person to have a right-of-way over one's land is an example of this type of easement. Such an easement is not assignable nor capable of inheritance. It should be given in writing with a time stipulation. Care should be taken that the use of the land is an easement, so that the use is not subject to adverse possession (squatter's rights). The person granted the easement can only use the land for the use granted.

When a party wall is built with mutual consent along a property line, each owner not only owns the wall on his property, but also, each owner has an easement for that part of the wall on his neighbor's land. The wall cannot be removed without consent of both parties. The easement transfers with the title to the land.

An easement may be obtained by prescription when a landowner builds a structure on his land that extends over (usually by mistake) onto his neighbor's land. If the owner of the encroached-upon land does not object within six years of the encroachment, the right to force the movement of the structure is lost in accordance with state statutes.

2. *License.* A license is a privilege granted to use a person's land for a stated purpose. It is usually given orally and may be terminated at any one time at the pleasure of the landowner. Without the granting of the use license, the subject use of the land by the nonlandowner would be trespass. For example, a private university has roads passing over its land and it closes those roads for 24 hours once a year (say, Christmas Day) to maintain private ownership of that land and private control of the roads.

3. *Profit à Prendre.* This French phrase means "profit to take." A person given a *profit à prendre* has the right to take from the land a product of value. Most state statutes require such a right in writing and such a document must be carefully prepared. Rights to minerals or wells may fall in this classification. Title remains in the hands of the landowner.

E. EMINENT DOMAIN

This is the right of a public agency to purchase land from an individual for a public purpose when the holder of title to the land is not willing to sell. No individual has absolute ownership of land in the United States. A fee simple title is the strongest interest a person, persons, or business entity can have in land. However, upon proper application and decree from the courts, a public agency has the right to condemn this land and make purchase. A court will not grant this right of eminent domain unless the public agency shows the intended use of the land is in the public interest. Without this right, the government would find it impossible to construct facilities that are needed for the welfare of the people. The building of a highway or railroad could be stopped by just one landowner refusing to sell a piece of land. Without eminent domain, schools, fire stations, etc., could not be located in the proper position.

The state law sets forth the proper procedure in the acquisition and payment of land to be taken by eminent domain. Landowners should not be approached with regard to the selling of their land until it is certain what land is needed. The true market value should be set by qualified assessors. Legal procedures prescribe the right of appeal from the first offered price. A procedure that is common is to name a set of independent assessors to establish a price when the government agency's price is rejected. As a last resort, the landowner can appeal the price to the courts. He will need special evidence that the price was not correct as given by the assessors before a judge will increase the sale price of the property. The landowner will have to pay any legal fees if he appeals the sale price to the courts.

Because of inflated land values, there have been financial hardships imposed on homeowners through loss of their land and homes under eminent domain. This occurs when a home and land is taken in a lower level economic area. The owner may obtain a fair market value, but he is unable to purchase another home of like quality in the same neighborhood or even in the same city. In some states, the law has been changed to require a government agency to not just give the owner the market price, but when so requested, the government agency must find and purchase a home of similar or better quality and make a trade. This eliminates the need for the homeowner to have to purchase more expensive property and acquire more debt when he does not desire to do so, or where he may have difficulty in obtaining a mortgage for a more expensive house. Fair compensation for a home may be a somewhat different opinion between an owner and an assessor. Here is another case where ethical considerations, on both the part of a homeowner and the persons in a governmental agency, should be the basis for an agreement. Eminent domain is a necessary right of government, but it should be used honestly and fairly.

F. THE EXPERT WITNESS

An area of legal jurisprudence in which an engineer or architect may become involved is the role of an expert witness. Civil cases involving liability and an ever increasing activity of environmental suits require the testimony of individuals knowledgeable in special fields of technology. A professional will be selected to present testimony because of his qualifications and reputation in the contested area of litigation. Both the plantiff's and defendant's legal counsel will most likely bring expert witnesses into the case.

An expert witness may be asked to only present written testimony or he may appear in court and take the witness stand to testify and be cross-examined by the opposition attorney. Expert witnesses are hired by the legal counsel of either the plaintiff or the defendant. The expert witness can charge whatever fee he desires but it should be within reason and at the generally accepted fee. The usual fee for the time in court is higher than the fee for just giving information or advice to legal counsel.

An expert witness should be well qualified by academic training and experience to answer questions that the lawyers for both sides will present. The legal firm which brings the particular witness into the case will try to select the person with the best qualifications that is available. A person is not required to accept a role as an expert witness when requested by a lawyer. However, a person can be subpoenaed by the court to appear at the trial as an expert witness. However, it is rare that lawyers request this action since they cannot have a chance to talk with the witness prior to his appearance in court. The lawyer who initiates the subpoena cannot be sure the statements of the expert witness will strengthen his case. If there is a degree of uncertainty on what an expert witness may say, the lawyers will probably prefer to not have such a person testify. When an expert witness appears because of a court order, he is paid the standard fee that has been established by law and he is paid by the court. This fee is usually considerably less than a law firm will pay for the services of an expert witness.

Because the expert witness when engaged by a lawyer is paid a professional fee, he may feel an obligation to strengthen the case of his lawyer client. However, he should be careful to only present the correct and true technical information as his knowledge provides him. He should be careful about giving opinions for which he does not have backup data. His testimony will be challenged by the opposing lawyer who may also be getting advice from another expert witness. The opposing lawyer will try by all legal means to discredit the testimony of an expert witness. He may even challenge the qualifications of a witness. If the witness has ever published conflicting opinions on a subject at two different times, the opposing lawyer will, if he is aware of this, present this as testimony. In being fair to all parties in the litigation, as well as protecting one's reputation, an expert witness should always maintain a high ethical standard throughout the case. Honesty and truthfulness should always be the background for any evidence and testimony. To knowingly lie is contrary to the oath taken on the witness stand and can lead to a charge of perjury. Conviction of perjury can lead to a jail sentence and/or fine. The ASCE Code of Ethics in statement 3.c states, "Engineers, when serving as expert witnesses, shall express an engineering opinion only when it is founded upon adequate knowledge of the facts, upon a background of technical competence, and upon honest conviction."

Answering the questions of a trial lawyer should be done in a careful deliberate manner. A trial lawyer may want to resort to "tricks" by calling for yes or no answers to his questions. A witness is not required to answer just yes or no if such an answer is not in keeping with the question. An opposition lawyer may also resort to leading questions in order to confuse or trap the witness into agreeing to a condition that did not exist.

A lawyer should thoroughly prepare his witness in all the facts and details of the case. The witness should do the necessary homework in preparation for his testimony. The witness should always visit the site of the subject litigation if the case involves a specific location. The expert witness must be very familiar with all the literature on the subject area in which he claims to be an expert. An opposing lawyer may cite references that will contradict the statements of the witness. This is especially true when the case involves environmental problems. Lack of knowledge of what other experts have written will discredit a witness before a judge or jury. An expert witness should be present throughout the entire trial so that he is aware of all previous testimony.

An expert witness is not permitted to cross-examine a witness of the opposing party. This appears to be a weak point in the legal process. Cross-examination has to be done by registered lawyers. An expert witness can advise a lawyer on what questions to ask a witness. Cross-examination is subject to rules of evidence which must be adhered to in court. Lawyers are trained in the knowledge of these rules and in the art of evidence presentation. Despite questions or procedures that might produce the "truth," specific court rules must be followed. When a witness gives evidence at a trial, all statements are recorded and made a record of the court proceedings, so as to be available at any future time. Answers to questions should be clearly and slowly given and in as simple language as possible.

G. PATENTS

Civil engineers and architects are not often involved with patent matters. Mechanical, electrical, and chemical engineers are much more likely to have concern with patents. Involvement with patents can be through two avenues. One avenue is the obtaining of a patent on a device or process invented by an individual or group. The second is the use of a device or process that has a patent. Both of these two involvements with patents will be discussed in this section.

A patent is a grant of right to exclude others from the making, using, or selling of an invention during a specified time. A patent is a legal monopoly. The legal basis of patent rights is for the purpose of protection to an invention. It is also for the purpose of promoting the progress of science and the arts. It secures for a limited period of time to authors and inventors exclusive rights to their respective writings and inventions. The Constitution of the United States recognizes this right. Incentives for new inventions would be very limited without patent laws and the granting of rights to creative individuals.

The patent system began in 1790. The system is administered by the Federal Government through the Patent and Trademark Office.

A person can obtain a patent by placing an application with the U.S. Patent Office. The requirements for patent applications are very exact and a person filling out an application should probably seek expert help in the preparation of it. An error in preparation may jeopardize the success of obtaining a patent. A search is necessary to determine whether the device or idea has already been patented. It must be shown that the subject of the patent is original and does not infringe upon any other patent. A statement to that effect is required. Drawings and a detailed description of the subject must also be part of the patent application.

If two people apply for a similar patent independently, the Patent Office will conduct a proceeding to determine who developed the work first. An inventor should keep a dated log of his work and have the log signed by a trusted witness. Such dated logs can be strong evidence in a patent hearing to determine who is the original inventor.

When an inventor receives a patent, he is granted exclusive rights to that patent for a period of 17 years. After this period of time, the patent becomes public property and anyone may use, manufacture, or sell the device. Upon approval of a patent, it is recorded and placed on file in the Patent Office. It then becomes public knowledge. Anyone with a prospective patent can (and must) search the patent office files to determine if his device has already been patented. Many registered patents never develop into a successfully marketed product.

A patent that is judged sensitive to national security will not have public disclosure. Federal government regulations will control the manufacture and use of such inventions.

The government will not police the infringement of any patent. This is the responsibility of the patent holder. Courts consider patents in the same manner as contracts. A patent holder can sue for damage from a person or organization using, manufacturing, or selling a patented product without permission of the patent holder. Anyone making an agreement of manufacture or sale with a patent holder should be sure that a valid contract is signed with the patent holder. Terms, including proper consideration, should be clearly defined.

Patent ownership can be transferred through an estate by a properly drawn will. The 17 years remains as the total life of the patent.

If an inventor wishes to manufacture and market the product during the period of patent application, he may do so, provided he is sure there is no infringement on an already granted patent. If manufacture begins before patent approval, infringement by others may be prevented by labeling the product with the words "patent pending."

An engineer may be required to sign a patent ownership document when he goes to work with a company. He may not be required to sign a patent policy but may be given a copy of the company policy. This then legally binds him to the company rights for any inventions in which he may be involved while in the employment of the company.

Company patent rights policies vary widely. Some company patent rights agreements will require all proceeds from the marketing or use of the patent to be the property of the company. Other companies may have a policy of dividing monetary proceeds. A patent is given in the name of an individual or individuals not in the name of a company or partnership. However, the inventor can assign to an employer the right to use that patent. The degree of required rights in the use of a patent by an employer will usually depend upon whichever following conditions apply.

1. The employee developed the patent in the course of his daily work activity for which time he was paid a salary by his employer.
2. The employee developed the patent independently of his employment duties but used company facilities, the work being done during "off-hours."
3. The work was done away from employment and was devoid of the use of company facilities, but the idea came from or was related to employment.
4. The patent is entirely exclusive from employment with regard to time, facilities, and relationship.

Company patent policies may cover, with different degrees of rights, all of the above situations. A person who is of an inventive mind should make a study of a company's patent policy and agreement before selecting employment. Such agreements are legally binding on the employee while employed and may be binding for a period of time after leaving the employment of a company.

A copyright is legally similar to a patent. It is a protection by federal statute or common law giving authors and artists exclusive right to the publication of their creative work. Copyrights are for a period of 28 years, with a 28-year renewal.

The details of copyright law change with time. New problems have developed in recent years with regard to copyright law. These new problems have been the result of automatic copy machines. The making of copies of any copyrighted material for sale is contrary to the law. The copying of a page or two for personal use is not an infringement of a person's copyright but copying a whole book is an infringement. Copying of materials for use in the education process has special rules. A ruling on copyright infringement would be based primarily on what extent the reproduction of material would have on the market value of the copyrighted work and if there would be a financial loss to the holder of the copyright because of the copying action.

Copyrights can be simply obtained by filing the necessary forms, paying a nominal fee, and sending a copy of the copyrighted material to the Library of Congress.

13. TORT AND LIABILITY

A. INTRODUCTION

A decade or two ago the word liability was threatening to only the most affluent; however, in the complex, interrelating world of the last half of the 20th century, the word has fearful overtones to a great many people. In a society that is lawsuit happy and looking for quick riches, liability can be a threat to people of even modest means.

The automobile has introduced the threat of liability to anyone who drives or owns one. It is mandatory almost everywhere in the United States and many other countries that a car owner carry liability insurance so that funds are available to pay the "injured" person who is not responsible for the accident in which he is involved. Injury can be to a person or property and is usually both. Liability suits involving automobiles have become so numerous and burdensome to the courts that new types of insurance have been required by law.

Whenever someone causes harm to the body or property of another person, he can and mostly likely will be asked to pay for the damages. Over the years, common law in the United States and some other countries has developed to such an extent that if it can be proven that damage has been suffered by one party because of the actions of another party, the person at fault quite likely will be required to make restitution to the owner of the damaged property. An area where this previous statement is untrue is in the case of criminal damage. The law does not really exempt the criminal from legal claims of liability. The criminal most often has little wealth to pay any damage claims so a legal suit is not entered. Nevertheless, in recent years a few such claims have been entered against criminals. A crime, although perpetrated in many cases against individuals, is usually classified as an offense against the public for which a governmental authority takes issue with the criminal and his act.

People in any walk of life can be brought to court for the purpose of paying a damage claim to the alleged injured party. Some examples are: A person who slips on another person's icy sidewalk may enter a legal suit and collect damages. A neighbor's child climbs up a ladder you have leaning against a tree and falls and is injured. The neighbor sues and will most likely collect. Your dog bites the mailman or paper boy. You most likely will settle out of court if the damage claim is modest. If the claim is extreme, you or your insurance company may end up hiring a lawyer and spending time, money, and anxiety.

The courts are full of legal cases of lawsuits with one party suing another because of acts of commission or omission which the plaintiff claims has done damage to his person or property. Court cases result when the plaintiff and defendant do not agree on the responsibility for the damage and if there is damage, what is the just and fair monetary payment for that damage?

The courts when requested have the responsibility for fixing the blame for damage and the value to the injured for the damage. This has proven to be a most difficult task. In a high percentage of cases, it would take more than the wisdom of Solomon to provide a just and fair answer. In many cases legal maneuvering and skill and play of emotions determines the outcome of a case more than justice and fair play.

When Jane Doe sues Bill Brown and wins, it it not likely that Bill Brown pays the bill or even an appreciable amount of the judgment. If Bill Brown has insurance to cover his judgment, thousands of people, possibly including you and me, pay Jane Doe for the damage done to her by Bill Brown. For example, if a plaintiff sues the Ford Motor Company and collects nearly a million dollars, then all the stockholders indirectly pay the plaintiff, and even those who buy the automobiles pay when car prices increase due to legal judgments or increased insurance costs. When a doctor is assessed a judgment, the results touch many doctors because of their liability insurance. Liability insurance premiums are paid by the patient's medical bills.

The various kinds of liability insurance protect thousands of people from financial disaster, yet it also invites thousands of claims that would otherwise not be made. Insurance in its basic operation is a means of inducing thousands of people to each share a small amount in the loss to a person or property that a party has suffered because of some negligent action. This may appear as a charitable action, but in reality most do this not from a feeling of sharing the hardship but strictly to protect one's own assets.

The remainder of this chapter will discuss tort liability and its various applications to the work of an engineer/architect. Of course, it is not possible to develop this subject in detail or in depth in just a few pages—whole books have been written on the subject. Law students spend many days in the study of this subject. The objective herein is to present to the engineer/architect (and primarily the student) the basic concept of tort liability and how it applies in his profession.

B. DEFINITION OF TORT

A tort is a civil wrongful act done by one party to another for which the victim can legally demand redress. A civil wrong is a wrong that is not a criminal act. The parties involved in torts may be individuals, partnerships, trusts, corporations, or any legally constituted organization. It is possible for a single act to be classified as both a tort and a crime. It was mentioned in Section A that some suits have been entered in the law courts where the offender was charged with a crime and also became a defendant in a damage claim.

A tort can be committed either intentionally or unintentionally. It can be due to an act of commission or omission. Tort acts of omission may be more common among engineer/architects.

Torts are classified into personal and property torts depending on the nature of the damage.

In tort suits, the defendants may be a single entity or several. Whether a plaintiff claims damage as the result of an action (or inaction) of several parties, he can bring suit against each in individual cases or they can all be named joint defendants in a single case. However, the plaintiff can collect only once for the same injury. The plaintiff cannot sue one party for an injury (personal or property) and in winning that suit, collect damages and then institute the same suit against another responsible party. The usual pattern in liability suits, when more than one party is involved in the action, is to name all the parties as codefendants in one suit.

C. TORT LIABILITY

In order to win a tort liability suit in a court of law, the plaintiff must establish five points. These points are essential prerequisites to a successful negligence suit. They are as follows:

1. A duty existed to use proper care and attention in the situation in question.
2. The defendant did not exercise the proper care and diligence which would be considered standard under the circumstances.
3. There was a close relationship between the lack of care and damage caused the plaintiff.
4. The defendant had no defense in his action or lack of action.
5. The plaintiff suffered damage because of the action.

Each of these five points will be discussed in detail and examples given.

If any one of these prerequisites is lacking or cannot be proven in a court of law, then it is very unlikely that the plaintiff can win the case. If one or more of the points has limited evidence, then the success of the suit will depend upon the skill of the plaintiff's lawyer and who is suing who. For example, a poor widow suing a large corporation in a jury trial would have a greater chance of success on a considerably weaker set of evidence than one corporation contesting another. This shows that the human element is present in any court case, and irrespective of feelings as to what is or is not justice, there will always be this lack of certainty in the courts. All people have biases and weaknesses in varying degrees and these human frailties can influence juries and even judges.

D. THE DUTY OF A PROFESSIONAL

A judgment cannot be rendered against a party unless that party had a duty to perform. A simple example would be that party A was being accosted on the street by a man with a gun. Party B witnessed the holdup and did not even go to a phone and call the police. Party A may want to sue Party B for his lack of diligence, but he has no case, since it has not been established by the law that a witness to a crime has a duty to notify the police.

A motorist has an accident while driving over a rough detour at a highway construction project. He not only sues the highway contractor, but also the design professional. The injured party may have a good case against the contractor because the detour road is his responsibility. However, the design engineer has no responsibility for the detour road and any suit against the designer would most likely end in failure for the plaintiff.

What is the duty of a professional, to what fine points does this duty extend and for what length of time, are questions that have been argued in the courts for many years and likely will be a major point in many court cases for many years in the future. A major difficulty in finding answers to these questions is that each project is different in some respect from any other project. The professionals involved in engineering projects have a wide range of duties that may vary greatly even on similar projects. A great contributing factor to the uncertainty of duty in a court trial is that the people having to make the final decision are the judge and/or the jury. Such nontechnical people will have to sift the conflicting statements of the trial lawyers and witnesses and reach a verdict.

In a recent case in New York, a workman employed by a subcontractor sued the general contractor for damages he suffered when he fell from a ladder erected in violation of the state's labor law. The ladder was erected by the subcontractor. The suit claimed the general contractor was negligent and also disobeyed the labor law. The trial court disallowed the negligence claim but did rule in favor of the plaintiff on the labor law claim. However, the Appellate Division of the New York State Supreme Court reversed the ruling and dismissed the laborer's suit. The court ruled that the general contractor had no duty to perform since they were not required to give direction as to the manner or method in the performance of the work of erecting the ladder.

An interesting court case that has not been tried at the time of this writing is the damage claims of the John Hancock Mutual Life Insurance Company against the architect, general contractor, glass supplier, and two bonding companies for damages sustained during construction of its 60-story building in Boston. A large amount of the glass in the windows cracked and all of the 10,344 panes had to be replaced with a new window pane design. The damage suit will probably involve several million dollars. The suit alleges that the architect, contractor, and glass supplier were all negligent in that they had a duty to perform and failed to perform all the necessary parts of that duty. It is quite likely that all defendants will attempt to show that they did perform whatever duties they had with regard to the windows. The general contractor may take the defense that he had no duty to perform with regard to design or installation. The results of the litigation will be very interesting from a legal standpoint. After all the data is gathered including extensive research on the problem, the case may still be settled out of court.

E. PERFORMANCE OF PROFESSIONAL DUTY

Under the law almost everyone has a duty to perform. If one is issued a driver's license by the state, the driver has a duty to operate an automobile in a safe manner and in accordance with the laws of the land. A homeowner must maintain safe premises to prevent injury by anyone passing through. If he has a swimming pool, neighborhood children must be protected from falling in and drowning. As a person becomes more involved with activities, his duties increase with regard to maintaining safe premises.

An engineer/architect, by the nature of his work, has many duties of a very special nature. He may be involved in the production of products or facilities that will be used by many people. The facilities or products with which he contributes his time and talents may cost hundreds of thousands or millions of dollars. If anything does not function properly or there is a failure of any element, the damage may be very great. Cost in time and money could be monumental. An engineer/architect has a special responsibility to perform his professional duties.

The question in many cases of litigation is what is acceptable or unacceptable performance, and who is to judge this? In a case that goes to litigation, the judge or jury must decide what is the required level of performance and if the defendant met that required level. Both the defendant and plaintiff will have legal help which in theory will attempt to obtain the answer to the question of required level of performance and whether the required performance had been met or a duty breached. Legal counsel will present evidence by documentation and witnesses. The burden of proof rests with the plaintiff in United States courts to prove a duty or a breach in the performance. It will also be necessary for the plaintiff to prove that the duty was owed to him.

Someone cannot sue unless they themselves have been harmed. Such harm does not have to be property or bodily harm to the plaintiff directly, but can be indirect. A suit can be entered if the harm is done to a minor under the plaintiff's care or to a person who supports the plaintiff who thereby loses that support.

When a person or any legal entity fails to perform his duty, the failure to perform is called negligence. A court (judge) will certainly dismiss a tort liability case if there is no evidence of negligence. A person may suffer damage and there still may be no case of negligence except the negligence of the person harmed. Structures have collapsed and killed or injured people, or at least caused thousands of dollars worth of damage and there was not sufficient evidence of negligence on the part of anyone connected with the design and construction of the structure. Some such failures have been due to abnormal loading conditions such as earthquakes, extreme snow loads, wind loads, or high water. Some physical phenomenon may happen that was previously unknown to the profession. The wind oscillations of the Tacoma Narrows Bridge was a case in point. The bridge engineering profession was almost completely unaware that a bridge designed with such physical dimensions and characteristics would vibrate in such a wild manner. Nobody was legally accused of negligence. Today, if an engineer designed a suspension bridge and it responded to wind in such a way that it caused the destruction of the bridge, the engineer would be declared negligent in his duties. Who would say he was negligent? His peers would certainly pass such a judgment if called (as they would be) to appear as expert witnesses. Through model testing and other research, design tools are available to determine the likely response of a bridge structure to wind or earthquake forces.

During the period of development of the airplane from the Wright Brothers until now, there have been many airplane crashes. Only a small percentage have resulted in tort liability suits. The standard of care that would have been required to prevent such losses was beyond the knowledge of the human race at the time of many of the accidents. Knowledge of fatigue of metals was nonexistent during the early years of flying. However, today such knowledge is common among structural and mechanical engineers and lack of that knowledge or failure to apply that knowledge would be classed as negligence.

Even though the required degree of necessary knowledge to prevent a loss is beyond the defendant's ability, it will not excuse negligence today. A design professional should seek whatever degree of professional help is necessary to adequately design a structure. This would be expected of him in a court of law. In the design of large complex facilities such as a bridge, dam, power plant, airplane, etc., many specialists are necessary. In today's modern age, engineering is too complex for any one person to have total requisite knowledge. The person or persons responsible for a design must know the required degree of expertise and engage the necessary consultants for a safe design. This is now the required duty of a professional.

Students in the author's structural design classes always ask the question, "How close is one required to follow the design specifications?" If an engineer is designing a building in a city that has a building code, then he must follow that code very carefully for two reasons. First, the city building department will not likely approve the plans if the code has not been followed in detail. Second, if there are any problems with the finished structure and the design professional has deviated from the code, then he is suspect and may be included in the liability suit even though the fault is not due to any design weakness. Quite likely, the design professional will, in any case of failure or distress, be included as a defendant along with all the others that have had any responsibility in the design and building of the facility. The plaintiff's lawyer will seize upon any evidence that the design professional did not follow the code or rules of accepted procedure. The designer will then be forced into the position of defending his action. Even if he brought into the design the skills and knowledge above and beyond what is represented by the code, he will have to prove this to a jury composed of people who have little knowledge of this world's physical laws beyond the law of gravity. This does not mean that a design engineer shouldn't allow a small percentage of overstress in a structural element in order to arrive at a readily available size. No reasonable expert witness would fail to admit that the science of structural mechanics or knowledge of strutural loadings are that precise as to forbid some leeway. However, any deviation from design codes or established standards of practice should have sound justification. New research to justify deviant procedures must have been or is in the process of being accepted by the profession.

In summary, it can be said that the required performance of an engineer/architect is that which is considered the norm for the profession. Keeping abreast of new developments, standards, and research is a must for any engineer or architect. This is one of the best liability insurance programs a professional of any type can have.

F. THE RELATIONSHIP BETWEEN DAMAGE AND NEGLIGENCE

In order for a plaintiff to win a tort liability suit, he must prove that the damage he received was directly related to a breach of duty required of the defendant. The conditions "perhaps" or "possibly" should not be accepted by the courts. This does not mean they haven't been used to try to win a court case. A defendant's lawyer should be alert to the lack of substantial evidence linking damage and negligence. As an example, a bridge collapses under a vehicle whose weight was below the posted load limit. The driver was seriously injured. Police records show that a truck driver was given a traffic ticket two weeks earlier for driving over the bridge with a heavier than posted load. The injured driver of the vehicle that was on the bridge when it collapsed decides to sue the owner of the truck claiming that the overloaded truck was responsible for weakening the structure so that the bridge later collapsed. The suit claimed a breach of duty on the part of the truck driver in not obeying the posted load limit on the bridge.

Although the structure collapsed, it would be difficult to prove that the first truck's passage on the bridge was the "straw that broke the camel's back." There could have been many other vehicles passing over the bridge of even heavier weight before or after the initial truck named in the suit. Technically, it would be very difficult to prove that the negligence of the truck driver was directly related to the damage caused to the plaintiff.

An example of a different verdict is the case of a new highway construction. A detour around a section of new highway construction involved a rather sharp turn. A vehicle coming upon the detour turnoff failed to negotiate the turn and rolled over, injuring the driver and several occupants. The injured parties filed suit against the state highway department and the road contractor. Negligence was claimed on the basis that detour warning signs were not posted far enough in advance of the detour entrance. The defense claimed that the accident was due not to faulty warning signs but to lack of vehicle control on the part of the driver. This type of case would require expert witnesses from the engineering profession to give testimony on the adequacy of the sign to prevent *any* driver traveling within the posted speed limit from having an accident. If it could be proven that the driver was traveling within the speed limit, then it is quite likely that a jury would return a verdict against the defendants. Of course, there would have to be no adverse testimony against the driver for any violations such as speeding, drunken driving, driving without a license, and so on.

In the case of a damage claim based on faulty engineering or architecture, the plaintiff would have to prove that inadequate performance was directly the cause of the damage. Lack of subsoil testing could not be claimed as the negligent action for the collapse of a structural beam when all calculations showed the beam to be adequate to carry the loads and that there was no abnormal settlement of the foundation. The plaintiff would have to determine another reason for the beam failure, and possibly would have to charge the contractor or material supplier and not the design professional.

G. DEFENSE

In all liability cases the defendant will have to decide, with the advice of legal counsel, whether he has a defense to the charges that have been brought against him. If the decision is that he has no defense or from a legal standpoint his chance in the courtroom is not good, he will try to reach a settlement out of court. This will save him the legal costs (which might be considerable) as well as a possible lower damage claim. Any such decision will, of course, have to be agreeable to the insurance company if the defendant has liability insurance.

If upon the advice of the liability insurance company and legal counsel that the defendant has a defense, then proceedings toward litigation will take place. This action may follow even though the defense feels it may suffer adverse court judgment, but believes the damage claims are excessive and the award to the plaintiff by the court will be less than an out-of-court settlement.

There may be mitigating circumstances for the action or lack of action that is being charged as negligence. In 1978, a court case was settled in a dispute between a contractor and a material supplier. A supplier had a contract to supply limestone to a construction contract but before any stone was delivered, the owner cancelled the use of limestone in the project. The contractor then cancelled the contract with the supplier of limestone. The contract between the owner and the general contractor permitted the owner to cancel or modify the contract. The contract showed estimated quantities of limestone but stated that, "Payment will be made only for actual quantities of work completed."

The supplier, although having delivered no limestone, sued the contractor for his loss in profit. The trial court ruled in favor of the supplier of limestone. However, the circuit court of appeals reversed the decision, ruling that the contract was not breached because it was a requirements contract, and since the owner removed the requirement for limestone there was none required. The court ruled that the supplier should have been knowledgeable with regards to the contract between the owner and contractor and the changes and terminations provisions of that contract. This brings forth an important point for a subcontractor to be aware of. The provisions of the contract between the owner and general contractor may apply in a contract between the general contractor and a subcontractor even though not specifically spelled out in the subcontract. It would be prudent for the subcontractor to assume this to be the case unless provisions in the subcontract stated a contrary condition. Ignorance of this legal precedence cost the limestone supplier considerable legal fees for which he received nothing in return.

H. DAMAGE TO THE PLAINTIFF

In order to collect on a liability suit, the plaintiff must prove damage. As stated in a previous section, this damage must be to a person or property.

The damage must be proven to be the result of the action. The damage to a person or property must not have occurred due to any previous action. The defendant's lawyer should make an investigation into whether any of the damage was present before the alleged damage took place. Medical histories will be investigated and attending physicians may be called as witnesses.

The value of the damage to the plaintiff will be determined by the court. If the damage is property damage, then third parties who are knowledgeable in costs will be asked for an estimate. If a facility or any part of it is damaged, the court will most likely require that the facility be returned to its predamaged condition and that the defendant pay the cost of restoration. A good example of this is an automobile accident. The defendant driver, or his insurance company, will have to restore the other vehicle to its preaccident condition if he is judged at fault.

A portion of a block wall around a baseball field fell over in a windstorm. The winds were within the limit range required by the city building code. Expert testimony supplied by the owner of the wall showed that the design was not adequate to withstand the wind force. The wall lacked sufficient reinforcing steel. The architect who designed the facility was found negligent and was required to pay for the wall replacement, incorporating adequate reinforcing steel, as well as paying the cost of the additional construction to adequately brace the portion of the wall that did not collapse. The architect was also required to pay court costs.

When a defendant is found guilty of negligence in a civil suit, the costs assessed by the court may be more than just the cost of repair. In a case in a midwestern state, a contractor had a contract to construct storm and sanitary sewers and waterwork improvements for a city. A separate contract to another contractor was let for building the pavement over the water mains. Later the pavement cracked badly and an investigation proved that the fill in the water and sewer line trenches, which supported the sidewalk, had not been compacted to the degree required in the sewer and waterline contract. The trial court awarded $58,400 to the city which was increased to $518,000 by a higher court.

The contractor appealed the ruling to the state supreme court after the necessary repair costs proved to be about $80,000. The contractor contended that the city would receive a windfall "profit" of over $400,000. The supreme court rejected the argument of the contractor ruling that the amount of money above the cost of repairs did not constitute a prohibited windfall.

In cases of proven fraud, the defendant can be assessed punitive damages, but this is not usually the case in a civil suit involving tort liability.

However, damages can be more than the repair or displacement cost of a damaged facility. It may include court costs and legal and incidental costs of the plaintiff. Incidental costs may also include cost of inconvenience. If the tort is a breach of contract, attorney's fees are recoverable by the plaintiff only if the contract so states or if statute authorizes the recovery of attorney fees.

When damages are to a person, the costs can be whatever the court considers reasonable. The recent trends in bodily damage awards have been increasing at a fast rate. A person disabled for life or a long period of time will usually receive much more than just lost wages. The award will usually include an amount to cover pain, suffering, and trauma. Awards against large corporations tend to be much higher than against individuals or small businesses.

There have been damage claims covering a very wide range of "injury" such as emotional distress, psychic distress, shock not due to injury, miscarriage due to emotional impact, etc. Some courts have awarded damage claims for such indirect damage. However, in general the courts have been reluctant to award claims for this type of damage. In a case where a mother suffered severe emotional shock when she witnessed a fatal accident involving her child, even though she was not a part of the accident, the court awarded damages to the mother.

Courts have granted awards to wives and/or children for the loss of companionship of a husband or father. This is in addition to the loss of financial support. It is of course difficult to establish the economic value of such a loss of companionship. There are no established values and an award would be at the "whim" of the court. The plaintiff may seek an award from the defendant that matches or comes close to the limit of the liability insurance of the defendant. This is especially so if the defendant has limited assets. In such cases, the insurance company will shoulder most of the defense since it is their funds that will be lost if the court case is lost.

Some personal damages that do not involve personal injury are recoverable. These are awards that are granted in libel and slander cases and they will be covered in a later section.

The trend to award very large (almost exorbitant) claims in liability suits has caused a great deal of concern among the professional community and insurance companies. Insurance premiums have almost become prohibitive in some types of liability insurance. This has been especially true in the medical profession. What was considered the norm of the court in previous years with regard to damage awards is no longer the present situation. Public concern has been provoked, since "overpaid" claims are, in reality, passed on to the consuming public in higher medical fees, higher insurance rates, higher bids for engineered facilities, etc. Discussion of some actions with regard to limit of liability is in a following section.

The legal pattern for damage awards has been that the more wrongful the conduct, the larger the award and the greater the likelihood of recovery against the defendant. A fraudulent action will always create a liability and a negligent action will most likely do the same. Gross negligence, which is stated as a disregard for the health, safety or rights of others, will almost always precipitate liability action if damage results from this negligence.

Contributory negligence is negligence on the part of the plaintiff himself which contributed to his own damage. Such negligence on the part of the plaintiff, even though slight, will in almost all cases result in loss of recovery. A common type of contributory negligence is where a road is clearly marked as closed and a driver nevertheless goes down the road and has an accident. Since the driver was negligent, he would not likely recover any damages. However, contractors should be very careful that all construction sites are well marked and the public is adequately protected from injury by unknowingly entering into the construction area.

Signs in themselves are not sufficient enough to warn a person or to prevent injury when children might trespass upon a construction site. Since children may not be able to read or may lack a sense of danger, construction sites, should be fenced if at all possible. This is usually a mandatory requirement of a liability insurance company.

If there is a requirement (as is now the case) that all people on a construction site must wear hardhats, and a person is injured because he neglects to do so, he will most likely lose a liability suit.

I. LIABILITY OF LANDOWNERS

The act of owning land requires responsibility. Lack of responsibility (duty to perform) can lead to a charge of negligence. Such a charge can result in a damage claim.

There have been sufficient lawsuits and judgments against landowners that the required duties of a landowner are quite clearly defined. People entering on land have the right to be protected from hazards due to the fault of the owner of the land. Liability for harm depends upon the circumstances which defines the duty of the landowner toward the plaintiff. An example was given in the previous section of a motorist ignoring posted signs and driving down a closed road. The court is very likely to judge that the posting of adequate signs is a duty.

A contractor is responsible for the entire construction site while the contract is in force. Even though the contractor has the major responsibility for the safety of the premises, the owner may be and has been named in liability suits when damage occurs to a third party. An owner should insist that the contractor eliminate hazards immediately. A clause to that effect should be part of the contract. OSHA regulations are for the purpose of eliminating hazards at work sites. Failure to comply with OSHA regulations is a breach of the law. Disregard for the law is likely to be considered gross negligence in a court. A contractor should be aware of any duty as prescribed by law. The owner should likewise be diligent and insist upon full compliance with all laws and good safety practices.

Any property owner has duties such as keeping sidewalks, stairways, etc. free from snow, ice, and slippery or hazardous objects. Unless the property is fenced and secured, anyone may come upon the property and they must be reasonably protected. Most cities have an ordinance requiring the property owner to keep the sidewalk in front of their property free of any hazard even though the city usually is the owner of the sidewalk. A person slipping, falling, and sustaining injury could claim negligence on the part of the landowner. It may be difficult to define what is a reasonable duty in clearing a sidewalk of snow. Requiring that a sidewalk be cleared by 6 a.m. after it snowed at night would seem unreasonable, but a safe sidewalk by noon would very likely be judged as a required performance of duty.

The question of duty to protect trespassers on a person's land is an unresolved legal problem. In the past, trespassers were generally considered breakers of the law and, therefore, landowners had no duty to perform with regard to trespassers. However, there have been rulings recently that have placed responsibility upon the landowner to protect *anyone* from a hazardous condition. This is especially so when the hazard is man-made or unnatural. A well-posted property can help in eliminating liability of a property owner to trespassers.

Special rules have been made for trespassing children. In most jurisdictions, there is a liability to the owner or leasee of the land for injuries to trespassing children. This is especially the case when an artificial condition exists on the land. The responsibility of protecting children (even though uninvited) from the hazards of a swimming pool is well known. Anything that attracts children and may be hazardous establishes the responsibility of the possessor of the land to protect children from the hazard. Such items as ladders, scaffolding, construction equipment, and many other items fall into this category.

Not only has the contractor been named in liability suits but also the owner and design professional when death or injury occurred on a construction site. The design professional has been declared responsible when he has had responsibility to the project during construction. Resident engineers and their employers have frequently been named as co-defendants in construction site personal injury claims.

J. AVOIDING LIABILITY SUITS

It has been rightfully stated, from past statistics, that an engineer/architect practicing as a design professional will be faced with at least one liability claim in his career. Large professional firms will have several such suits in the period of a decade or two. Developing procedures to avoid such is a reasonable and profitable pursuit. Some rules have been formulated to reduce the probability of contending with a plaintiff. Some of the most noteworthy are listed below.

1. Be thorough in your engineering/architectural work. Hire competent people and be sure supervision is adequate and qualified.
2. Keep records of all transactions with client and contractor. Decisions affecting the design and construction should be in writing. Minutes of conferences should be kept and initialed by all parties.
3. The design professional should not be involved in any decisions relating to construction procedure or policies unless that is part of the contract with the client. Any indication that the design professional is involved in the construction process will open the door to his being named as a defendant, along with the contractor and owner, in an injury or damage claim. Stay away from periodic visits to the construction site if you have no responsibility in the area.
4. Keep the area where your employees are working as free of hazards as possible. Be sure all equipment, including automobiles, are in proper working order. Repair defects promptly. Do

not let employees or anyone else operate equipment unless they have been instructed in its use. Institute safety rules in the use of any equipment.

5. Obey all laws and be firm with employees that they do likewise. If your firm is involved in construction be sure that all OSHA laws are known and adhered to.

6. Make complete, neat design computations and be sure they are checked and initialed by the checker. Keep these together with all drawings for the record. They may be needed 10 or 15 years in the future.

7. Follow all codes pertaining to the design of any facility. Do not deviate from code requirements unless there is a good reason, and document such changes. Exceed minimum standards where possible, without adding to the cost of the facility.

8. If the work you are called to perform is beyond your experience or expertise, then refer the job to other firms experienced in that area or involve consultants who are proven experts.

9. Make sure all your work is at a level equal to or above the standards of the profession.

10. All technical employees should keep current in the profession by enrolling in refresher or continuing education courses, attending technical conferences, belonging to and maintaining activity in technical societies, and developing highly competent abilities.

11. Do not depend on exculpatory contract clauses to shift your liability exposure to the owner or builder. Such dodges may receive scant recognition in court. Posting of signs declaring a party is not liable does not remove the possibility of legal claims. A party can always be declared liable if negligence and duty to perform can be proven.

12. If your firm is involved in inspection of others' work, be sure your inspectors are qualified and thorough. Remind them that they work for you, not the contractor. Have them report to you any hazardous conditions on the job site. Relay this information to the owner and the contractor by memo.

K. IMMUNITY FROM LIABILITY

In the past, most government agencies and their employees were immune from tort liability and legal action for breach of contract. Employees were immune only while performing their official duties. The federal government still carries this immunity to a lesser extent but can be sued if granted permission from Congress.

The situation with regard to state and local governments varies from state to state. There are a variety of statutes relating to who may be sued. In recent years the trend has been to remove immunity for states and municipal governments.

Some states have protected charitable organizations from liability suits. The law has been changing away from this immunity. Sovereign and charitable immunity have been created by the courts but legislatures have been removing such procedures. There has been less public pressure to remove sovereign immunity when a government body is acting in a governmental function such as police protection, fighting fires, supplying of culinary water, etc. However, a government unit may not have immunity from damage claims when they were acting in a capacity frequently performed by private organizations. Such activities as operation of sports facilities, parking lots, etc. would fall in this latter category.

In 1887 Congress consented, by passage of the Tucker Act, to suits against the federal government for breach of contract. Thus, the Court of Claims was established.

Claims against government units usually must strictly follow prescribed procedures. There are usually shorter time limits for starting legal actions against public agencies than against private parties.

Public agencies may have sovereign immunity with regard to the performance of design or construction contracts. This immunity may or may not extend to the private design professional or private contractor. A private organization may want to determine the extent, if any, of immunity they have when under contract to a government agency.

L. STATUTES OF LIMITATIONS

Statutes of Limitations is any law which fixes the time within which parties must take judicial action to enforce rights or else thereafter be barred from enforcing them. The rationale behind such laws is the belief that if a long period of time has lapsed since the action took place permitting a legal claim, the evidence would become stale. Another reason is the belief that a possible defendant should eventually be relieved of any concern over a legal claim. Assembling sound, accurate evidence after a long period of time is usually very difficult.

In statute of limitations laws the beginning of the time period may vary among legal jurisdictions and the type of legal claim. The length of time to bring action will vary considerably depending upon the subject matter of the litigation. The time period may be extended for bringing action when a plaintiff is incapacitated. If the defendant moved outside of the jurisdiction of the court, the legal time period for commencing the suit could also be extended. It should be kept in mind that valid legal claims may be lost if legal action is not commenced within a time period specified by law.

Most recently some design professionals have begun to be sued for work done by them many years previously. Some court jurisdictions have ruled that the work carried no statute of limitations. For example, a geotechnical consulting firm was recently sued for work performed 10 years ago. The settlement of some large chemical storage tanks was greater than that predicted by the consulting engineering firm. In this case the judge stated the jury should consider that soils engineering is not an exact science and that the law does not require the engineer to have an infallible judgment. Nevertheless, the jury returned a verdict against the soils engineering firm. An appeal is pending.

In a 1978 case in California, a Court of Appeals upheld the constitutionality of the state's statute of limitations protecting engineers. The state statute is four years from substantial completion of the facility. The claim was for an auto accident that injured a plaintiff in 1964. The plaintiff claimed that the intersection as designed by the engineering firm prior to 1964 was faulty and was the cause of the accident. The court rejected the plaintiff's claim that the statute of limitations violated the equal protection clause of the constitution.

A Wisconsin Court of Appeals rejected the claim by legal counsel for an architect that the statute of limitations barred a negligence suit against his client. The charges were that the architect who designed an apartment complex was negligent in the design of the roof. The project was completed in July 1970. In April 1977 the owner brought suit claiming that the roof had begun to leak and had been rotting for the previous four years. The state statute of limitations was six years for suits involving damage to property. The architect's attorney argued that the suit was nine months too late.

Both the trial and appellate court refused to dismiss the case citing that the statute of limitations begins when the damage occurs (the roof started leaking in 1973) and not when the construction was completed. In this case the court refused to set the beginning date of the statute of limitations at the time of negligence, even though Wisconsin has a six-year special statute of limitations which begins to run at substantial completion.

Professional groups may want to closely scrutinize the wording of state statute of limitations laws to clarify the beginning date of such laws. There appears to be a wide variation of court rulings on this aspect of law.

M. LIABILITY OF AN EMPLOYEE

There is not an active history of liability claims against employees of engineer/architect firms. Most suits are against a company and not directed at an individual professional unless he is in business for himself. One of the primary reasons for this is that an employee is not likely to have large assets or liability insurance to pay any claim which may be assessed against him. The company is a more fruitful target. A jury case against a firm will have a much greater chance for success than would a case against an individual employee.

Laws vary from state to state on the legal actions possible against employees by either the employer or a third party. In the past, a loyal employee with high standards of ethics and conduct was protected and shielded by the employer and was also covered by the company liability insurance policy. The present-day situation may be somewhat changed because of the great number of professional employees some firms now have and also because of the unionization of these employees. In either of these two cases the employee may be quite removed from top management. The feeling of protecting and shielding an employee may not be strong when this feeling of distance between employee and management exists.

It may be possible for a company to sue an employee for damages when the company suffers great loss because of negligent action on the part of the employee. When this negligent action or incompetent action occurs, the usual action of the employer is to dismiss the employee. A suit to recover any damages caused by the employee is not likely to meet with success. Recovery from an employee in a successful suit would in most cases be meager because of the minimal assets of an employee. The legal costs would probably be greater than any possible recovery. The employer would suffer some loss in public relations as well as some ill feeling among other employees.

When an employee commits an act of fraud, embezzlement, or gross misconduct, an employer can sue for damages as well as bring criminal charges against the employee. Intracompany crime is difficult to prosecute since law enforcement with white-collar crimes has not been as diligently pursued as with public crime. If such cases are brought to court, the employer may face a charge of defamation of character if the employee is not proven guilty.

One area of employee activity where the engineer/architect should be knowledgeable of the legal consequences is that of reporting any criminal action of an employer. If the company is found guilty of a criminal act, an employee who was knowledgeable of this action, even though not involved, could conceivably be punished for lack of disclosure. An employee who commits a criminal act under direction from an employer is nevertheless guilty of the crime.

An officer of a company could be criminally charged and face lawsuits for civil damages for criminal acts of his company even though he personally did not take part in the action, but was knowledgeable and failed to report the act to legal authorities. Falsifying of time records by an engineering firm with a cost-plus contract would be a criminal act that could bring criminal charges against any employee making a false record, or against any officer of the company.

N. LIABILITY SUITS AGAINST DESIGN FIRMS AND CONTRACTORS

A sample of negligence suits against engineer/architect firms and contractors is presented herein. The defendant will not be named in the legal citations given. The purpose is to describe the types of action that have brought on litigation and not to study the legal details. Reference to the court case is also unnecessary for the purpose.

1. In 1972 a boy was riding on a raft in a pool of water which had collected behind a flooded and totally submerged culvert. The suction from the culvert pulled the boy into the connecting drain pipe and he drowned.

The boy's father filed suit against the local sewer district, the contractor, and the design engineers, claiming a defect in design and construction. The trial court ruled that the boy's negligence contributed to the accident. The case was appealed and the appellate court reversed the ruling but ruled that the design firm was the only defendant since the sewer district enjoyed sovereign immunity and the statute of limitations protected the contractor. The state of jurisdiction had no special statute of limitations for lawsuits against architects and engineers.

Evidence was produced by the plaintiff at the lower court's trial that the culvert pipe as designed was too small, the design unsafe, and the catch basin did not comply with accepted design standards. The opinion of the appellate court was that evidence of negligence on the part of the engineers was present and they authorized a full trial. The trial is pending at the time of this writing.

2. Shortly after a school building was built in Arkansas, the roof began to leak. Numerous attempts for repair were made over a period of seven years, with little permanent effect. After this period of time the architect declared to the owner the need for replacing the roof. The owner then filed suit against the general contractor, the roofing subcontractor, and the architect charging all of them with negligence and breach of warranties. The owner hired another roofing contractor to replace the roof for $155,000.

The trial court ruled in favor of all the defendants. The owner appealed to the Supreme Court of Arkansas which ordered the defendants to stand trial. At the trial the Supreme Court ruled that the five-year statute of limitations protected the architect.

3. A large concrete cooling tower in New Jersey collapsed during construction, killing 51 workers. Claims have been made that the concrete as placed did not meet specifications and that there were inadequate tests on the concrete in the process of construction. Legal proceedings, which undoubtedly will be numerous and extensive, have not entered the court. This case will no doubt involve large sums of money. When such a large and extensive project is in process, care and diligence on the part of all parties is of utmost importance.

4. A contractor sued a consulting firm for damages resulting from flooding at a construction site. The contractor contended that there was flood damage to a partially completed sewage treatment plant. The flood required the completed work to be completely demolished and rebuilt.

The state supreme court ruled that the design engineer for the facility had failed to properly design flood control dikes to protect the construction during periods of high water. According to the court, such negligence was the direct cause of the damage and the contractor was entitled to recover the cost of reconstruction and expense of flood cleanup.

5. A concrete viaduct in a South American country collapsed during construction, killing 29 persons and injuring many others. The engineer in charge of the design was given a suspended jail sentence of one year, four months, and ten days. It was ruled that he authorized openings in the concrete deck to remove forms. Such openings were not in the original plans. The quality of concrete was also questioned.

The incarceration of design professionals has taken place in some countries when gross negligence has been proven. Such is very rare in the United States, but such extreme action is legal when fraud or deliberate negligence can be proven as the cause of personal injury or death.

6. In many kinds of underground construction, methane gas may develop in excavations. Such gas is present where organic soils exist. When explosions take place because of the presence of underground gas, the engineer/architect may be liable if he did not indicate in the plans and specifications that gas may be present and to also indicate the ventilation facilities that would be required to prevent explosions.

Broken gas lines have resulted from improper backfill and compaction around the pipes. If the engineer/architect is responsible for inspection, explosions due to negligent inspection can result in damage claims against the contractor, subcontractor, and engineering firm responsible for inspection. An award for over a million dollars was granted to a plaintiff when a father was killed and three others in the family suffered severe burns because of a gas explosion due to a break in a gas line. Testimony indicated frozen material had been placed in a backfill and the work was haphazard.

Construction of utility lines, especially gas lines, should be carefully performed. Poor-quality workmanship can cause severe damage and be costly to all involved in the construction.

7. A new purchaser of a nine-year-old building found bad cracks in the brick walls. A special mortar additive was used in the construction. The additive had been approved by all parties responsible for the construction. The owner claimed that excessive steel rust on steel members in contact with the brick was due to the mortar additive.

A suit for five million dollars was recently directed against nine companies involved in the erection or material supply during the construction. Considerable testing will be required to prove or disapprove the owner's charges. There will no doubt be much litigation to prove guilt or innocence of negligence. Statute of limitations may also come into the case when it finally develops.

8. A contractor was sued by the widow of a workman killed in the construction of the Houston Astrodome. The workman was employed by a painting subcontractor. The suit was directed against the general painting contractor who had subcontracted a portion of the work to the deceased employee's company. The whole case hinged on who was responsible for not erecting a safety net. The court ruled that the negligence was the fault of the general contractor and not the subcontractor. The court did not find the subcontractor liable to the contractor or to the widow.

Responsibility of safety measures between contractor and subcontractors can many times be undefined by specifications, contract clauses, or regulations. Lack of proper attention to such matters can result in death or serious injury and costly court settlements. Absence of proper safety measures while parties dispute responsibility is a very foolish procedure.

9. A contractor was held liable for a subcontractor's employee by the U.S. Court of Appeals. Although the OSHA rules required the subcontractor to support a trench, the contractor also had a duty to the subcontractor's employees. The contract provisions required the contractor to comply with OSHA rules and thus by extension required the contractor to provide a safe working place for subcontractor's employees. A subcontractor's employee was injured in a trench cave-in. The contract clause made the contractor liable as well as the subcontractor.

O. LIBEL AND SLANDER

Libel is a written or published statement, picture, etc. that is likely to harm the reputation of the person about whom it is made. Slander is defined as a false report meant to do harm to the good name and reputation of another. Both slander and libel are meant to accomplish the same result, i.e., to damage the reputation of another. Slander is by the medium of the spoken word and libel is by the written word or photograph. Both are tort acts and damages can be collected in a court if a plaintiff can prove the defendant guilty and damage done.

If what is written (libel) can be proven to be untruthful and intended to defame a person, no proof of damage is necessary. The law will infer that third persons have read the statements and damage has been committed. However, in slander cases, it must be proven that the plaintiff has suffered damage as a result of the statement. A case of slander where damage does not actually have to be proven is when a defendent makes a statement that is detrimental to the plaintiff's business. Slanderous statements pertaining to an act of moral turpitude or unchastity is slander per se. Intention of defamation may be hard to prove, but some statements in themselves convey such an intention. Statements with regard to the honesty or ability of one's professional competitor would most likely be considered as a purposeful attempt at defamation.

There are some defenses against being charged for libel or slander. The strongest defense is truth. It has been stated that when truth is the defense, the motive is immaterial to the case. A person cannot be sued for slander if as a witness he makes a truthful statement in court. If it is proven he made false statements a suit for slander will most likely be successful as well as a charge of perjury.

Other defenses that might mitigate any damages because of defamatory statements are a prompt retraction of the statement, the honest belief of the defendant in the truth of his statement, and the bad reputation of a plaintiff. Introducing evidence to support the latter may be very difficult in the courtroom.

False statements by parties to a contract have been subject to damage claims in the court. One such case was the result of a supplier of materials for a subcontractor communicating to the prime contractor and owner that he had not been paid by the subcontractor. Proof that the communication was false was grounds for a damage claim.

A design professional must be careful of any statements made to the news media when an engineering failure of any nature takes place. Opinions should be clearly stated as opinions and should not be lightly made or made in a manner that could be construed as intended defamation. Dependence on a defense of some nature may lead one to take risks in unsubstantiated statements. The defense may be sufficient to defend against a judgment but not sufficient to defend against a court appearance. Libel and slander are very reprehensible breaches of ethical codes and will most likely be treated as such by peers and technical societies.

P. ACTION TO CONTROL LIABILITY LITIGATION

The rash of legal claims and high monetary awards that have been granted by the courts in recent years is causing great distress among professionals and contractors. It is also of major concern of many thinking people in the business community. Exorbitant awards are very disruptive to the economic climate of society and have resulted in people taking the attitude that one should take advantage of every opportunity to gain all one can out of any situation disregarding standards of what is right or wrong. Society is the loser when the moral fiber of a nation decreases.

There is little fault with the belief that one should be responsible for his actions and penalties should follow for wrong behavior. This belief has been a part of society since at least the time of Hammurabi and Moses. There is also a long held belief that punishment should fit the crime.

The present court system leaves much to be desired when applied to the solution of liability and malpractice suits. Such litigation becomes involved with complex debate of subjects of a highly technical nature. A lay jury must sift the facts presented by highly specialized and highly trained expert witnesses, who may themselves not agree on whether the operation performed by a doctor or the structural design executed by an engineer was done correctly. It is not reasonable to assume that correct and fair judgments will always come from such a jury system. It is very possible that a jury in civil trials will only arrive at the correct judgment by happenstance. This is not to state that juries are not honest or do not seek a fair and correct judgment. It is almost too much to ask of people who are almost always entirely unqualified people to arrive at a decision that is based upon information that is on the whole beyond their understanding.

Such a system breeds ills among the legal profession. If the jury is unable to comprehend the complexities of the case, contending attorneys will then deduce that the decision of the jury will be based on who are the "good guys" and the "bad guys". The attorney will then assume his role to be that of proving to the jury who they are. Standards of ethical and fair dealing can then become quite strained in such a legal climate. In a case involving a large business entity and a "poor widow," it is rather easy for a lawyer to convince a jury who is the "bad guy" and who is the "good guy". In such a legal environment, jury decisions are disproportionately biased in favor of the relative skills in the courtroom of the contending attorneys. When this happens the jury decision is influenced less by the facts and ethics of the case.

The legal community as well as the technical and medical professions have recognized the weakness of the system of jury trials for civil action suits involving liability and malpractice. A system is available that may result in a better judgment. This system is the use of a judge and not a jury. It has been used in some cases. However, either the plaintiff or the defendant can demand a jury trial. Most plaintiffs prefer a jury trial if the plaintiff is an individual versus a business entity as the defendant.

It is very possible that a judge is no more qualified to judge the facts of a case involving a highly technical activity than a jury. However, the astuteness of a judge who has many years of legal experience would qualify him to be the better judge of the legal merits of a case as well as better able to ignore emotional appeals from the attorneys. Personal biases against one party or another may be an influence in both a bench judgment trial or a trial by jury.

Some countries have used a multiple-judge system—usually amounting to three judges. This has the merits of a judge over a jury, especially if the judges are experienced in civil cases involving technical or professional subjects. The multiple-judge system minimizes the effects of the personal bias of a single person.

A newly proposed procedure that shows considerable merit and is gaining approval in professional areas has been put forth by the U.S. Department of Justice. This is a screening procedure which would apply to all medical malpractice lawsuits. Such lawsuits would be first screened by a board of experts who would review each case and judge the merit of each claim. The opinion of the panel would not prevent a case from being taken to court but the findings of the board would discourage such a procedure if the plaintiff's charges were judged to have little merit. The findings of the board could be entered into the court proceedings.

If the review board did find that malpractice had taken place, then the defendant and his liability insurance company would most likely attempt an out-of-court settlement. Regardless of the ruling of the review board, there would no doubt be a considerable reduction in the caseload of the courts. The review-board procedure would help to modify the injustice of "shotgun" suits wherein a plaintiff names everyone that has been connected with the project in his lawsuit. In some cases the rationale is not that all defending parties were guilty of a breach of duty and thus all contributed to the damage, but if many parties divide the damage claim among them, then the amount may not be very large. A modest amount from each one will equal the value of the damage claimed and will be sufficiently small so that an out-of-court settlement would appear to be the best answer for all. A court case for any one defendant may produce legal fees equal to or greater than the out-of-court settlement. There is then no sense in risking a court case unless a defendant believes strongly in his innocence and also that the jury will return a not-guilty decision.

The "shotgun" approach to liability claims is a form of "legal blackmail" that is not uncommon in damage claims occurring in the course of construction of large facilities. The review-board screening method would help in reducing this type of liability claim.

Other procedures have been suggested for bringing greater justice to liability and malpractice lawsuits. Upper limits of liability claims, require the plaintiff to pay all legal and time expense fees of the defendant if the defendant wins the case, signing release from liability before any work begins, and other ideas have been suggested. To date little has been done, although state legislatures have studied and considered the problem. The problem is complex and remedial action is necessary for the well being of the country. The objective in any remedial plan would be to seek justice for all parties.

BIBLIOGRAPHY

In addition to the references given at the end of the chapters, other publications on the subject of this book that may be of interest to the reader are listed here.

1. L. D. Simpson and E. R. Dillavou, *Law for Engineers and Architects*, 4th ed., St. Paul, Minn.: West Publishing Co., 1958.
2. *What Everyone Needs to Know About Law*, U.S. News and World Report Books, 1971.
3. C. W. Dunham, R. D. Young, and J. T. Bockrath, *Contracts, Specifications, and Law for Engineers*, 3rd ed., N.Y.: McGraw-Hill Book Co., 1979.
4. N. Walker, E. N. Walker, and T. K. Rohdenburg, *Legal Pitfalls in Architecture and Building Construction*, 2nd ed. N.Y.: McGraw-Hill Book Co., 1979.
5. M. Stokes, *Construction Law in Contractor's Language*. N.Y.: McGraw-Hill Book Co., 1977.
6. *Ethics, Professionalism, and Maintaining Competence*, Reprints of an ASCE Speciality Conference, March 1977.
7. The American Society of Civil Engineers has a periodic publication titled *Engineering Issues*. The following titles treat the subject of this book:
   Engineering Issues—January 1972
   "Integrity and the Ecological Crisis," by Myron Tribus.
   "Ethical and Social Responsibility in the Planning and Design of Engineering Projects," by David Mann.
   "Professional Development for Consulting Engineers," by Gerald L. Baker.
   Engineering Issues—October 1974
   "Professionalization—and a Relevant Code of Ethics," by Charles R. Schrader.
   Engineering Issues—January 1975
   "Motivating Student Towards Professional Thinking," by Joseph M. DeSalva.
   "A Statement on Professional Services—Competitive Bidding," by Gerald Devlin.

"Public Versus Client Interests—An Ethical Dilemma for the Engineer," by Terry L. Turnick.
Engineering Issues—July 1975
"Social Responsibility within the Present Code of Ethics," by George Govatos.
"Kickback Versus Professional Ethics," by Russel C. Jones.
Engineering Issues—January 1976
"When is a Political Contribution Influence Buying?" by Thomas J. McCollough.
Engineering Issues—April 1976
"Requirements of Professional Practice," by David Novick.
"Environmental Decision-Making and Facilities Citing," by Michael S. Baram.
Engineering Issues—July 1976
"Kansas Solution to Unionization of Professionals," by Bruce F. McCollum.
"Is There Price Collusion in the Civil Engineering Profession?" by Michael N. Goodkind.
Engineering Issues—April 1977
"The Story Behind the Recent National Scandals Involving Engineers," by Brian J. Lewis.
"Unionization of American Engineers," by Benjamin E. Burritt.
"Orientation Toward Professional Development" by Richard P. Long.
Engineering Issues—October 1978
"Ethics in Inter-Firm Cooperation," by Jack McMinn.
"New Challenges—New Liabilities," by Luther W. Graef.
"Examinations for Registration," by William J. Hanna.

8. M. Stone, "The Crime of Cheating," *U.S. News and World Report*, September 19, 1977, p. 92.

9. J. Barzun, "The Professions Under Seige," *Harpers*, October 1978.

10. *Legal Briefs* for Architects, Engineers, and Contractors, a twice-monthly newsletter published by McGraw–Hill.

11. R. E. Vansant, "Vansant's Law," *AE Concepts in Wood Design*, periodic publication of the American Wood Preservers Institute.

APPENDIX: CODES OF ETHICS

NATIONAL SOCIETY OF PROFESSIONAL ENGINEERS
Code of Ethics

PREAMBLE

The Engineer, to uphold and advance the honor and dignity of the engineering profession and in keeping with high standards of ethical conduct:
- Will be honest and impartial, and will serve with devotion his employer, his clients, and the public;
- Will strive to increase the competence and prestige of the engineering profession;
- Will use his knowledge and skill for the advancement of human welfare.

Section 1   The Engineer will be guided in all his professional relations by the highest standards of integrity, and will act in professional matters for each client or employer as a faithful agent or trustee.

a. He will be realistic and honest in all estimates, reports, statements, and testimony.

b. He will admit and accept his own errors when proven wrong and refrain from distorting or altering the facts in an attempt to justify his decision.

c. He will advise his client or employer when he believes a project will not be successful.

d. He will not accept outside employment to the detriment of his regular work or interest, or without the consent of his employer.

e. He will not attempt to attract an engineer from another employer by false or misleading pretenses.

f. He will not actively participate in strikes, picket lines, or other collective coercive action.

g. He will avoid any act tending to promote his own interest at the expense of the dignity and integrity of the profession.

Section 2   The Engineer will have proper regard for the safety, health, and welfare of the public in the performance of his professional duties. If his engineering judgment is overruled by nontechnical authority, he will clearly point out the consequences. He will notify the proper authority of any observed conditions which endanger public safety and health.

a. He will regard his duty to the public welfare as paramount.

b. He shall seek opportunities to be of constructive service in civic affairs and work for the advancement of the safety, health and well-being of his community.

c. He will not complete, sign, or seal plans and/or specifications that are not of a design safe to the public health and welfare and in conformity with accepted engineering standards. If the client or employer insists on such unprofessional conduct, he shall notify the proper authorities and withdraw from further service on the project.

Section 3 The Engineer will avoid all conduct or practice likely to discredit the profession or deceive the public.

a. The Engineer shall not make exaggerated, misleading, deceptive or false statements or claims about his professional qualifications, experience or performance in his brochures, correspondence, listings, advertisements or other public communications.

b. The above prohibitions include, but are not limited to, the use of statements containing a material misrepresentation of fact or omitting a material fact necessary to keep the statement from being misleading; statements intended or likely to create an unjustified expectation; statements containing prediction of future success; statements containing an opinion as to the quality of the Engineer's services; or statements intended or likely to attract clients by the use of showmanship, puffery, or self-laudation, including the use of slogans, jingles, or sensational language or format.

c. Consistent with the foregoing, the Engineer may advertise for recruitment of personnel.

d. Consistent with the foregoing, the Engineer may prepare articles for the lay or technical press. Such articles shall not imply credit to the author for work performed by others.

Section 4 The Engineer will endeavor to extend public knowledge and appreciation of engineering and its achievements and to protect the engineering profession from misrepresentation and misunderstanding.

a. He shall not issue statements, criticisms, or arguments on matters connected with public policy which are inspired or paid for by private interests, unless he indicates on whose behalf he is making the statement.

Section 5 The Engineer will express an opinion of an engineering subject only when founded on adequate knowledge and honest conviction.

a. The Engineer will insist on the use of facts in reference to an engineering project in a group discussion, public forum or publication of articles.

Section 6 The Engineer will undertake engineering assignments for which he will be responsible only when qualified by training or experience; and he will engage, or advise engaging, experts and specialists whenever the client's or employer's interests are best served by such service.

Section 7 The Engineer will not disclose confidential information concerning the business affairs or technical processes of any present or former client or employer without his consent.

a. While in the employ of others, he will not enter promotional efforts or negotiations for work or make arrangements for other employment as a principal or to practice in connection with a specific project for which he has gained particular and specialized knowledge without the consent of all interested parties.

Section 8 The Engineer will endeavor to avoid a conflict of interest with his employer or client, but when unavoidable, the Engineer shall fully disclose the circumstances to his employer or client.

a. The Engineer will inform his client or employer of any business connections, interests, or circumstances which may be deemed as influencing his judgment or the quality of his services to his client or employer.

b. When in public service as a member, advisor, or employee of a governmental body or department, an Engineer shall not participate in considerations or actions with respect to services provided by him or his organization in private engineering practice.

c. An Engineer shall not solicit or accept an engineering contract from a governmental body on which a principal or officer of his organization serves as a member.

Section 9   The Engineer will uphold the principle of appropriate and adequate compensation for those engaged in engineering work.

a. He will not accept remuneration from either an employee or employment agency for giving employment.

b. When hiring other engineers, he shall offer a salary according to the engineer's qualifications and the recognized standards in the particular geographical area.

c. If in sales employ, he will not offer, or give engineering consultation, or designs, or advice other than specifically applying to the equipment being sold.

Section 10   The Engineer will not accept compensation, financial or otherwise, from more than one interested party for the same service, or for services pertaining to the same work, unless there is full disclosure to and consent of all interested parties.

a. He will not accept financial or other considerations, including free engineering designs, from material or equipment suppliers for specifying their product.

b. He will not accept commissions or allowances, directly or indirectly, from contractors or other parties dealing with his clients or employer in connection with work for which he is responsible.

Section 11   The Engineer will not compete unfairly with another engineer by attempting to obtain employment or advancement or professional engagements by taking advantage of a salaried position, by criticizing other engineers, or by other improper or questionable methods.

a. The Engineer will not attempt to supplant another engineer in a particular employment after becoming aware that definite steps have been taken toward the other's employment.

b. He will not pay, or offer to pay, either directly or indirectly, any commission, political contribution, or a gift, or other consideration in order to secure work, exclusive of securing salaried positions through employment agencies.

c. An Engineer shall not request, propose, or accept a professional commission on a contingent basis under circumstances in which his professional judgment may be compromised, or when a contingency provision is used as a device for promoting or securing a professional commission.

d. While in a salaried position, he will accept part-time engineering work only at a salary not less than that recognized as standard in the area.

e. An Engineer will not use equipment, supplies, laboratory, or office facilities of his employer to carry on outside private practice without consent.

f. An Engineer will not use "free engineering" as a device to solicit or otherwise secure subsequent paid engineering assignments.

Section 12 The Engineer will not attempt to injure, maliciously or falsely, directly or indirectly, the professional reputation, prospects, practice or employment of another engineer, nor will he indiscriminately criticize another engineer's work. If he believes that another engineer is guilty of unethical or illegal practice, he shall present such information to the proper authority for action.

a. An Engineer in private practice will not review the work of another engineer for the same client, except with the knowledge of such engineer, or unless the connection of such engineer with the work has been terminated.

b. An Engineer in governmental, industrial or educational employ is entitled to review and evaluate the work of other engineers when so required by his employment duties.

c. An Engineer in sales or industrial employ is entitled to make engineering comparisons of his products with products by other suppliers.

Section 13 The Engineer will not associate with or allow the use of his name by an enterprise of questionable character, nor will he become professionally associated with engineers who do not conform to ethical practices, or with persons not legally qualified to render the professional services for which the association is intended.

a. He will conform with registration laws in his practice of engineering.

b. He will not use association with a nonengineer, a corporation, or partnership, as a "cloak" for unethical acts, but must accept personal responsibility for his professional acts.

Section 14 The Engineer will give credit for engineering work to those to whom credit is due, and will recognize the proprietary interests of others.

a. Whenever possible, he will name the person or persons who may be individually responsible for designs, inventions, writings, or other accomplishments.

b. When an Engineer uses designs supplied to him by a client, the designs remain the property of the client and should not be duplicated by the Engineer for others without express permission.

c. Before undertaking work for others in connection with which he may make improvements, plans, designs, inventions, or other records which may justify copyrights or patents, the Engineer should enter into a positive agreement regarding the ownership.

d. Designs, data, records, and notes made by an engineer and referring exclusively to his employer's work are his employer's property.

Section 15 The Engineer will cooperate in extending the effectiveness of the profession by interchanging information and experience with other engineers and students, and will endeavor to provide opportunity for the professional development and advancement of engineers under his supervision.

a. He will encourage his engineering employees' efforts to improve their education.

b. He will encourage engineering employees to attend and present papers at professional and technical society meetings.

c. He will urge his engineering employees to become registered at the earliest possible date.

d. He will assign a professional engineer duties of a nature to utilize his full training and experience, insofar as possible, and delegate lesser functions to subprofessionals or to technicians.

e. He will provide a prospective engineering employee with complete information on working conditions and his proposed status of employment, and after employment will keep him informed of any changes in them.

"By order of the United States District Court for the District of Columbia, former Section 11(c) of the NSPE Code of Ethics prohibiting competitive bidding, and all policy statements, opinions, rulings or other guidelines interpreting its scope, have been rescinded as unlawfully interfering with the legal right of engineers, protected under the antitrust laws, to provide price information to prospective clients; accordingly, nothing contained in the NSPE Code of Ethics, policy statements, opinions, rulings or other guidelines prohibits the submission of price quotations or competitive bids for engineering services at any time or in any amount."

STATEMENT BY NSPE EXECUTIVE COMMITTEE

In order to correct misunderstandings which have been indicated in some instances since the issuance of the Supreme Court decision and the entry of the Final Judgment, it is noted that in its decision of April 25, 1978, the Supreme Court of the United States declared: "The Sherman Act does not require competitive bidding."

It is further noted that as made clear in the Supreme Court decision:

1. Engineers and firms may individually refuse to bid for engineering services.

2. Clients are not required to seek bids for engineering services.

3. Federal, state, and local laws governing procedures to procure engineering services are not affected, and remain in full force and effect.

4. State societies and local chapters are free to actively and aggressively seek legislation for professional selection and negotiation procedures by public agencies.

5. State registration board rules of professional conduct, including rules prohibiting competitive bidding for engineering services, are not affected and remain in full force and effect. State registration boards with au-

thority to adopt rules of professional conduct may adopt rules governing procedures to obtain engineering services.

6. As noted by the Supreme Court, "nothing in the judgment prevents NSPE and its members from attempting to influence governmental action. . . ."

Note: In regard to the question of application of the Code to corporations vis-a-vis real persons, business form or type should not negate nor influence conformance of individuals to the Code. The Code deals with professional services, which services must be performed by real persons. Real persons in turn establish and implement policies within business structures. The Code is clearly written to apply to the Engineer and it is incumbent on a member of NSPE to endeavor to live up to its provisions. This applies to all pertinent sections of the Code.

NSPE Publication No. 1102 As revised, July 22, 1978

AMERICAN SOCIETY OF CIVIL ENGINEERS
Code of Ethics*
(Effective January 1, 1977)

FUNDAMENTAL PRINCIPLES**

Engineers uphold and advance the integrity, honor and dignity of the engineering profession by:

1. using their knowledge and skill for the enhancement of human welfare;
2. being honest and impartial and serving with fidelity the public, their employers and clients;
3. striving to increase the competence and prestige of the engineering profession; and
4. supporting the professional and technical societies of their disciplines.

FUNDAMENTAL CANONS

1. Engineers shall hold paramount the safety, health and welfare of the public in the performance of their professional duties.
2. Engineers shall perform services only in areas of their competence.
3. Engineers shall issue public statements only in an objective and truthful manner.
4. Engineers shall act in professional matters for each employer or client as faithful agents or trustees, and shall avoid conflicts of interest.
5. Engineers shall build their professional reputation on the merit of their service and shall not compete unfairly with others.
6. Engineers shall act in such a manner as to uphold and enhance the honor, integrity, and dignity of the engineering profession.
7. Engineers shall continue their professional development throughout their careers, and shall provide opportunities for the professional development of those engineers under their supervision.

---

*As adopted September 25, 1976.

Under the CODE OF ETHICS of the American Society of Civil Engineers, the submission of fee quotations for engineering services is not an unethical practice. ASCE is constrained from prohibiting or limiting this practice and such prohibition or limitation has been removed from the CODE OF ETHICS. However, the procurement of engineering services involves consideration of factors in addition to fee, and those factors should be evaluated carefully in securing professional services (Added July 1972).

**The American Society of Civil Engineers adopted THE FUNDAMENTAL PRINCIPLES of the ECPD Code of Ethics of Engineers as accepted by the Engineers' Council for Professional Development (ECPD). (By ASCE Board of Direction action April 12–14, 1975.)

ASCE GUIDELINES TO PRACTICE UNDER
THE FUNDAMENTAL CANONS OF ETHICS

1. Engineers shall hold paramount the safety, health and welfare of the public in the performance of their professional duties.
   a. Engineers shall recognize that the lives, safety, health and welfare of the general public are dependent upon engineering judgments, decisions and practices incorporated into structures, machines, products, processes and devices.
   b. Engineers shall approve or seal only those design documents, reviewed or prepared by them, which are determined to be safe for public health and welfare in conformity with accepted engineering standards.
   c. Engineers whose professional judgment is overruled under circumstances where the safety, health and welfare of the public are endangered shall inform their clients or employers of the possible consequences.
   d. Engineers who have knowledge or reason to believe that another person or firm may be in violation of any of the provisions of Canon 1 shall present such information to the proper authority in writing and shall cooperate with the proper authority. in furnishing such further information or assistance as may be required.
   e. Engineers should seek opportunities to be of constructive service in civic affairs and work for the advancement of the safety, health and well-being of their communities.
   f. Engineers should be committed to improving the environment to enhance the quality of life.
2. Engineers shall perform services only in areas of their competence.
   a. Engineers shall undertake to perform engineering assignments only when qualified by education or experience in the technical field of engineering involved.
   b. Engineers may accept an assignment requiring education or experience outside of their own fields of competence, provided their services are restricted to those phases of the project in which they are qualified. All other phases of such project shall be performed by qualified associates, consultants, or employees.
   c. Engineers shall not affix their signatures or seals to any engineering plan or document dealing with subject matter in which they lack competence by virtue of education or experience or to any such plan or document not reviewed or prepared under their supervisory control.
3. Engineers shall issue public statements only in an objective and truthful manner.
   a. Engineers should endeavor to extend the public knowledge of engineering, and shall not participate in the dissemination of untrue, unfair or exaggerated statements regarding engineering.
   b. Engineers shall be objective and truthful in professional reports, statements, or testimony. They shall include all relevant and pertinent information in such reports, statements, or testimony.

c. Engineers, when serving as expert witnesses, shall express an engineering opinion only when it is founded upon adequate knowledge of the facts, upon a background of technical competence, and upon honest conviction.

d. Engineers shall issue no statements, criticisms, or arguments on engineering matters which are inspired or paid for by interested parties, unless they indicate on whose behalf the statements are made.

e. Engineers shall be dignified and modest in explaining their work and merit, and will avoid any act tending to promote their own interests at the expense of the integrity, honor and dignity of the profession.

4. Engineers shall act in professional matters for each employer or client as faithful agents or trustees, and shall avoid conflicts of interest.

a. Engineers shall avoid all known or potential conflicts of interest with their employers or clients and shall promptly inform their employers or clients of any business association, interests, or circumstances which could influence their judgment or the quality of their services.

b. Engineers shall not accept compensation from more than one party for services on the same project, or for services pertaining to the same project, unless the circumstances are fully disclosed to and agreed to, by all interested parties.

c. Engineers shall not solicit or accept gratuities, directly or indirectly, from contractors, their agents, or other parties dealing with their clients or employers in connection with work for which they are responsible.

d. Engineers in public service as members, advisors, or employees of a governmental body or department shall not participate in considerations or actions with respect to services solicited or provided by them or their organization in private or public engineering practice.

e. Engineers shall advise their employers or clients when, as a result of their studies, they believe a project will not be successful.

f. Engineers shall not use confidential information coming to them in the course of their assignments as a means of making personal profit if such action is adverse to the interests of their clients, employers or the public.

g. Engineers shall not accept professional employment outside of their regular work or interest without the knowledge of their employers.

h. Engineers shall not review the work of other engineers for the same client except with the knowledge of such engineers, unless the assignments or contractual agreements for the work have been terminated. However, engineers in governmental, industrial or educational employment are entitled to review and evaluate the work of other engineers when so required by their duties.

5. Engineers shall build their professional reputation on the merit of their services and shall not compete unfairly with others.

a. Engineers shall not give, solicit or receive either directly or indirectly, any commission, political contribution, or a gift or other consideration in order to secure work, exclusive of securing salaried positions through employment agencies.

b. Engineers should negotiate contracts for professional services fairly and on the basis of demonstrated competence and qualifications for the type of professional service required.

c. Engineers shall not attempt to obtain, offer to undertake, or accept commissions for which they know other legally qualified individuals or firms have been selected or employed until they have evidence that the selection, employment or agreements of the latter have been terminated and they give the latter written or other equivalent notice that they are so doing.

d. Engineers shall not request, propose or accept professional commissions on a contingent basis under circumstances in which their professional judgments may be compromised.

e. Engineers shall not falsify or permit misrepresentation of their academic or professional qualifications or experience.

f. Engineers shall give proper credit for engineering work to those to whom credit is due, and recognize the proprietary interests of others. Whenever possible, they shall name the person or persons who may be responsible for designs, inventions, writings or other accomplishments.

g. Engineers may advertise professional services in a way that does not contain self-laudatory or misleading language or is in any other manner derogatory to the dignity of the profession. Examples of permissible advertising are as follows:

— Professional cards in recognized, dignified publications, and listings in rosters or directories published by responsible organizations, provided that the cards or listings are consistent in size and content and are in a section of the publication regularly devoted to such professional cards.

— Brochures which factually describe experience, facilities, personnel and capacity to render service, providing they are not misleading with respect to the engineer's participation in projects described.

— Display advertising in recognized dignified business and professional publications, providing it is factual, contains no laudatory expressions or implication and is not misleading with respect to the engineer's extent of participation in projects described.

— A statement of the engineer's names or the name of the firm and statement of the type of service posted on projects for which they render services.

— Preparation or authorization of descriptive articles for the lay or technical press, which are factual, dignified and free from laudatory implications. Such articles shall not imply anything more than direct participation in the project described.

— Permission by engineers for their names to be used in commercial advertisements, such as may be published by contractors, material suppliers, etc., only by means of a modest, dignified notation acknowledging the engineers' participation in the project described. Such permission shall not include public endorsement of proprietary products.

h. Engineers shall not maliciously or falsely, directly or indirectly, injure the professional reputation, prospects, practice or employment of another engineer or indiscriminately criticize another's work.

i. Engineers shall not use equipment, supplies, laboratory or office facilities of their employers to carry on outside private practice without the consent of their employers.

6. Engineers shall act in such a manner as to uphold and enhance the honor, integrity, and dignity of the engineering profession.

a. Engineers shall not knowingly act in a manner which will be derogatory to the honor, integrity or dignity of the engineering profession or knowingly engage in business or professional practices of a fraudulent, dishonest or unethical nature.

7. Engineers shall continue their professional development throughout their careers, and shall provide opportunities for the professional development of those engineers under their supervision.

a. Engineers should keep current in their specialty fields by engaging in professional practice, participating in continuing education courses, reading in the technical literature, and attending professional meetings and seminars.

b. Engineers should encourage their engineering employees to become registered at the earliest possible date.

c. Engineers should encourage engineering employees to attend and present papers at professional and technical society meetings.

d. Engineers shall uphold the principle of mutually satisfying relationships between employers and employees with respect to terms of employment including professional grade descriptions, salary ranges, and fringe benefits.

AMERICAN INSTITUTE OF CHEMICAL ENGINEERS
Code of Ethics

The Council adopted as the Code of Ethics of the American Institute of Chemical Engineers the 1962 Canons of Ethics of Engineers' Council for Professional Development.

FUNDAMENTAL PRINCIPLES OF PROFESSIONAL
ENGINEERING ETHICS

The Engineer, to uphold and advance the honor and dignity of the engineering profession and in keeping with high standards of ethical conduct:

    I. Will be honest and impartial, and will serve with devotion his employer, his clients, and the public;

   II. Will strive to increase the competence and prestige of the engineering profession;

  III. Will use his knowledge and skill for the advancement of human welfare.

RELATIONS WITH THE PUBLIC

1.1 The Engineer will have proper regard for the safety, health and welfare of the public in the performance of his professional duties.

1.2 He will endeavor to extend public knowledge and appreciation of engineering and its achievements, and will oppose any untrue, unsupported, or exaggerated statements regarding engineering.

1.3 He will be dignified and modest in explaining his work and merit, will ever uphold the honor and dignity of his profession, and will refrain from self-laudatory advertising.

1.4 He will express an opinion on an engineering subject only when it is founded on adequate knowledge and honest conviction.

1.5 He will preface any ex parte statements, criticisms, or arguments that he may issue by clearly indicating on whose behalf they are made.

RELATIONS WITH EMPLOYERS AND CLIENTS

2.1 The Engineer will act in professional matters as a faithful agent or trustee for each employer or client.

2.2 He will act fairly and justly toward vendors and contractors, and will not accept from vendors or contractors, any commissions or allowances, directly or indirectly.

2.3 He will inform his employer or client if he is financially interested in any vendor or contractor, or in any invention, machine, or apparatus, which is involved in a project or work of his employer or client. He will

248

not allow such interest to affect his decision regarding engineering services which he may be called upon to perform.

2.4 He will indicate to his employer or client the adverse consequences to be expected if his engineering judgment is over-ruled.

2.5 He will undertake only those engineering assignments for which he is qualified. He will engage or advise his employer or client to engage specialists and will cooperate with them whenever his employer's or client's interests are served best by such an arrangement.

2.6 He will not disclose information concerning the business affairs or technical processes of any present or former employer or client without his consent.

2.7 He will not accept compensation—financial or otherwise—from more than one party for the same service, or for other services pertaining to the same work, without the consent of all interested parties.

2.8 The employed engineer will engage in supplementary employment or consulting practice only with the consent of his employer.

RELATIONS WITH ENGINEERS

3.1 The Engineer will take care that credit for engineering work is given to those to whom credit is properly due.

3.2 He will provide a prospective engineering employee with complete information on working conditions and his proposed status of employment, and after employment will keep him informed of any changes in them.

3.3 He will uphold the principle of appropriate and adequate compensation for those engaged in engineering work, including those in subordinate capacities.

3.4 He will endeavor to provide opportunity for the professional development and advancement of engineers in his employ or under his supervision.

3.5 He will not injure maliciously the professional reputation, prospects, or practice of another engineer. However if he has proof that another engineer has been unethical, illegal, or unfair in his practice, he should so advise the proper authority.

3.6 He will not compete unfairly with another engineer.

3.7 He will not invite or submit price proposals for professional services, which require creative intellectual effort, on a basis that constitutes competition on price alone. Due regard should be given to all professional aspects of the engagement.

3.8 He will cooperate in advancing the engineering profession by interchanging information and experience with other engineers and students, and by contributing to public communication media, to the efforts of engineering and scientific societies and schools.

AMERICAN SOCIETY OF MECHANICAL ENGINEERS
Council Policy

ETHICS

ASME requires ethical practice by each of its members and has endorsed the following Code of Ethics of Engineers of the Engineers' Council for Professional Development as referenced in the ASME Constitution, Article C2.1.1.

Code of Ethics of Engineers

THE FUNDAMENTAL PRINCIPLES

Engineers uphold and advance the integrity, honor and dignity of the engineering profession by:

I. using their knowledge and skill for the enhancement of human welfare;

II. being honest and impartial, and serving with fidelity the public, their employers and clients;

III. striving to increase the competence and prestige of the engineering profession; and

IV. supporting the professional and technical societies of their disciplines.

THE FUNDAMENTAL CANONS

1. Engineers shall hold paramount the safety, health and welfare of the public in the performance of their professional duties.
2. Engineers shall perform services only in the areas of their competence.
3. Engineers shall issue public statements only in an objective and truthful manner.
4. Engineers shall act in professional matters for each employer or client as faithful agents or trustees, and shall avoid conflicts of interest.
5. Engineers shall build their professional reputation on the merit of their services and shall not compete unfairly with others.
6. Engineers shall associate only with reputable persons or organizations.
7. Engineers shall continue their professional development throughout their careers and shall provide opportunities for the professional development of those engineers under their supervision.

The original Canons were adopted by Engineers' Council for Professional Development, October 25, 1947 and accepted by The American Society of Mechanical Engineers the same year. There have been subsequent revisions. The latest version as presented above was approved by ECPD, October 1, 1974 and ratified by the ASME Council, March 16, 1975.

250

The ASME criteria for enforcement of the Canons are:

1. Engineers shall hold paramount the safety, health and welfare of the public in the performance of their professional duties.
   a. Engineers shall recognize that the lives, safety, health and welfare of the general public are dependent upon engineering judgments, decisions and practices incorporated into structures, machines, products, processes and devices.
   b. Engineers shall not approve or seal plans and/or specifications that are not of a design safe to the public health and welfare and in conformity with accepted engineering standards.
   c. Whenever the Engineers' professional judgment is over-ruled under circumstances where the safety, health, and welfare of the public are endangered, the Engineers shall inform their clients and/or employers of the possible consequences and notify other proper authority of the situation, as may be appropriate.
      c.1 Engineers shall do whatever possible to provide published standards, test codes, and quality control procedures that will enable the public to understand the degree of safety or life expectancy associated with the use of the designs, products, or systems for which they are responsible.
      c.2 Engineers shall conduct reviews of the safety and reliability of the designs, products, or systems for which they are responsible before giving their approval to the plans for the design.
      c.3 Whenever Engineers observe conditions which they believe will endanger public safety or health, they shall inform the proper authority of the situation.
   d. If engineers have knowledge or reason to believe that another person or firm may be in violation of any of the provisions of these Canons, they shall present such information to the proper authority in writing and shall cooperate with the proper authority in furnishing such further information or assistance as may be required.
      d.1 They shall advise the proper authority if an adequate review of the safety and reliability of the products or systems has not been made or when the design imposes hazards to the public through its use.
      d.2 They shall withhold approval of products or systems when changes or modifications are made which would affect adversely its performance insofar as safety and reliability are concerned.
2. Engineers shall perform services only in areas of their competence.
   a. Engineers shall undertake to perform engineering assignments only when qualified by education or experience in the specific technical field of engineering involved.
   b. Engineers may accept an assignment requiring education or experience outside of their own fields of competence, but their services shall be restricted to other phases of the project in which they are qualified. All other phases of such project shall be performed by qualified associates, consultants, or employees.

3. Engineers shall issue public statements only in an objective and truthful manner.
   a. Engineers shall endeavor to extend public knowledge, and to prevent misunderstandings of the achievements of engineering.
   b. Engineers shall be completely objective and truthful in all professional reports, statements or testimony. They shall include all relevant and pertinent information in such reports, statements, or testimony.
   c. Engineers, when serving as expert or technical witnesses before any court, commission, or other tribunal, shall express an engineering opinion only when it is founded upon adequate knowledge of the facts in issue, upon a background of technical competence in the subject matter, and upon honest conviction of the accuracy and propriety of their testimony.
   d. Engineers shall issue no statements, criticisms, or arguments on engineering matters which are inspired or paid for by an interested party, or parties, unless they preface their comments by identifying themselves, by disclosing the identities of the party or parties on whose behalf they are speaking, and by revealing the existence of any pecuniary interest they may have in matters under discussion.
   e. Engineers shall be dignified and modest in explaining their work and merit, and shall avoid any act tending to promote their own interest at the expense of the integrity, honor and dignity of the profession or another individual.
4. Engineers shall act in professional matters for each employer or client as faithful agents or trustees, and shall avoid conflicts of interest.
   a. Engineers shall avoid all known conflicts of interest with their employers or clients and shall promptly inform their employers or clients of any business association, interests, or circumstances which could influence their judgment or the quality of their services.
   b. Engineers shall not undertake any assignments which would knowingly create a potential conflict of interest between themselves and their clients or their employers.
   c. Engineers shall not accept compensation, financial or otherwise, from more than one party for services on the same project, or for services pertaining to the same project, unless the circumstances are fully disclosed to, and agreed to, by all interested parties.
   d. Engineers shall not solicit or accept financial or other valuable considerations, for specifying the products of material or equipment suppliers, without disclosure to their clients or employers.
   e. Engineers shall not solicit or accept gratuities, directly or indirectly, from contractors, their agents, or other parties dealing with their clients or employers in connection with work for which they are responsible.
   f. When in public service as members, advisors, or employees of a governmental body or department, Engineers shall not participate in

considerations or actions with respect to services provided by them or their organization(s) in private or product engineering practice.

g. Engineers shall not solicit an engineering contract from a governmental body on which a principal, officer, or employee of their organization serves as a member.

h. When, as a result of their studies, Engineers believe a project(s) will not be successful, they shall so advise their employer or client.

i. Engineers shall treat information coming to them in the course of their assignments as confidential, and shall not use such information as a means of making personal profit if such action is adverse to the interests of their clients, their employers, or the public.

    i.1 They will not disclose confidential information concerning the business affairs or technical processes of any present or former employer or client or bidder under evaluation, without his consent, unless required by law.

    i.2 They shall not reveal confidential information or finding of any commission or board of which they are members unless required by law.

    i.3 Designs supplied to Engineers by clients shall not be duplicated by the Engineers for others without the express permission of the client(s).

j. The Engineer shall act with fairness and justice to all parties when administering a construction (or other) contract.

k. Before undertaking work for others in which the Engineer may make improvements, plans, designs, inventions, or other records which may justify copyrights or patents, the Engineer shall enter into a positive agreement regarding the rights of respective parties.

l. Engineers shall admit and accept their own errors when proven wrong and refrain from distorting or altering the facts to justify their decisions.

m. Engineers shall not accept professional employment outside of their regular work or interest without the knowledge of their employers.

n. Engineers shall not attempt to attract an employee from another employer by false or misleading representations.

o. Engineers shall not review work of other Engineers except with the knowledge of such Engineers, or unless the assignments/or contractual agreements for the work have been terminated.

    o.1 Engineers in governmental, industrial, or educational employment shall review and evaluate the work of other engineers when so required by their duties.

    o.2 Engineers in sales or industrial employment shall make fair engineering comparisons of their products with products of other suppliers when required by their duties to make comparisons.

5. Engineers shall build their professional reputation on the merit of their services and shall not compete unfairly with others.

a. Engineers shall negotiate contracts for professional services on the basis of demonstrated competence and qualifications for the type of professional service required and at fair and reasonable prices.

b. Engineers shall not attempt to supplant other Engineers in a particular employment after becoming aware that definite steps have been taken toward the others' employment or after they have been employed.

c. Engineers shall not request, propose, or accept professional commissions on a contingent basis under circumstances under which their professional judgments may be compromised.

d. Engineers shall not falsify or permit misrepresentation of their, or their associates', academic or professional qualifications. They shall not misrepresent or exaggerate their degrees of responsibility in or for the subject matter of prior assignments. Brochures or other presentations incident to the solicitation of employment shall not misrepresent pertinent facts concerning employers, employees, associates, joint venturers, or their past accomplishments.

e. Engineers shall prepare articles for the lay or technical press which are only factual, dignified and free from ostentations or laudatory implications. Such articles shall not imply other than their direct participation in the work described unless credit is given to others for their share of the work.

f. Engineers shall not maliciously or falsely, directly or indirectly, injure the professional reputation, prospects, practice or employment of another engineer, nor shall they indiscriminately criticize another's work.

g. Engineers shall not use equipment, supplies, laboratory or office facilities of their employers to carry on outside private practice without consent.

6. Engineers shall associate only with reputable persons or organizations.

a. Engineers shall not knowingly associate with or permit the use of their names or firm names in business ventures by any person or firm which they know, or have reason to believe, are engaging in business or professional practices of a fraudulent or dishonest nature.

b. Engineers shall not use association with non-engineers, corporations, or partnerships as "cloaks" for unethical acts.

7. Engineers shall continue their professional development throughout their careers, and should provide opportunities for the professional development of those engineers under their supervision.

8. Any Engineer accepting membership in The American Society of Mechanical Engineers by this action agrees to abide by this Council Policy on Ethics and the procedures for implementation.

Responsibility: Policy Board, Professional and Public Affairs
Approved: March 7, 1976
Revised: December 9, 1976

INSTITUTE OF ELECTRICAL AND ELECTRONICS ENGINEERS
Code of Ethics

PREAMBLE

Engineers, scientists and technologists affect the quality of life for all people in our complex technological society. In the pursuit of their profession, therefore, it is vital that IEEE members conduct their work in an ethical manner so that they merit the confidence of colleagues, employers, clients and the public. This IEEE Code of Ethics represents such a standard of professional conduct for IEEE members in the discharge of their responsibilities to employers, to clients, to the community and to their colleagues in this Institute and other professional societies.

ARTICLE I

Members shall maintain high standards of diligence, creativity and productivity, and shall:

1. Accept responsibility for their actions;
2. Be honest and realistic in stating claims or estimates from available data;
3. Undertake technological tasks and accept responsibility only if qualified by training or experience, or after full disclosure to their employers or clients of pertinent qualifications;
4. Maintain their professional skills at the level of the state of the art, and recognize the importance of current events in their work;
5. Advance the integrity and prestige of the profession by practicing in a dignified manner and for adequate compensation.

ARTICLE II

Members shall, in their work:

1. Treat fairly all colleagues and co-workers, regardless of race, religion, sex, age or national origin;
2. Report, publish and disseminate freely information to others, subject to legal and proprietary restraints;
3. Encourage colleagues and co-workers to act in accord with this Code and support them when they do so;
4. Seek, accept and offer honest criticism of work, and properly credit the contributions of others;
5. Support and participate in the activities of their professional societies;
6. Assist colleagues and co-workers in their professional development.

255

ARTICLE III

Members shall, in their relations with employers and clients:

1. Act as faithful agents or trustees for their employers or clients in professional and business matters, provided such actions conform with other parts of this Code;
2. Keep information on the business affairs or technical processes of an employer or client in confidence while employed, and later, until such information is properly released, provided such actions conform with other parts of this Code;
3. Inform their employers, clients, professional societies or public agencies or private agencies of which they are members or to which they may make presentations, of any circumstances that could lead to a conflict of interest;
4. Neither give nor accept, directly or indirectly, any gift, payment or service of more than nominal value to or from those having business relationships with their employers or clients;
5. Assist and advise their employers or clients in anticipating the possible consequences, direct and indirect, immediate or remote, of the projects, work or plans of which they have knowledge.

ARTICLE IV

Members shall, in fulfilling their responsibilities to the community:

1. Protect the safety, health and welfare of the public and speak out against abuses in those areas affecting the public interest;
2. Contribute professional advice, as appropriate, to civic, charitable or other nonprofit organizations;
3. Seek to extend public knowledge and appreciation of the profession and its achievements.

Feb. 1979

Index

AAA, *see* American Arbitration
Association
AAES, *see* American Association of
Engineering Societies
ABA, *see* American Bar Association
Affirmative action programs, 90
Agency concept, 119-120
AIA, *see* American Institute of
Architects
American Arbitration Association
(AAA), 186-190
arbitration, commercial rules for,
187-188
mediation clause of, 187
American Association of Engineering
Societies, 10
American Bar Association (ABA),
20-23
code of professional responsibility,
20-21
American Institute of Architects (AIA)
Code of Ethics, U.S. Dept. of Justice
action against, 44-45
competitive bidding, educational
programs advising against, 45
contract forms available from, 136
American Society of Civil Engineers
(ASCE), 10-11, 16, 27, 34, 36
admonishment by, publication of, 56
antitrust violations of, 41-44
Code of Ethics of, 35-36, 123
business restraints of, 36
enforcement of, 39-40
expert witnesses, provisions for,
200

1977 revision of, 45
U.S. Dept. of Justice action
against, 41-46
Committee on Professional
Conduct (CPC) of, 39, 58-59
competitive bidding, educational
programs advising against,
43
employee-employer relations,
guidelines for, 47
engineer fee schedules available
from, 111
membership, suspension of, 54-55,
60
purpose of, 15-16
supplanting of other engineers,
prohibitions against, 58
Antitrust, *see* Sherman Antitrust Act;
U.S. Dept. of Justice
Appellate Courts, 91
Arbitration, 95-97
of contracts, 186-190
judgment
appeal of, 189
finality of, 188
Arbitrator, biases of, 96
ASCE, *see* American Society of Civil
Engineers
Attorneys
ethical conduct of, 20
prosecuting, 92
Authority
of agents, 119-120
of engineers, 146, 185-186
of inspectors, 146

employment positions of, 68–77,
110
environment, responsibility of, 35
in government, legal training
required of, 77
licensing of, *see* Registration of
engineers
in military, 74; *see also* U.S. Army
Corps of Engineers
moral responsibilities of, 125
public interest of, 13–14, 116
recruitment of, 47–48
registration of, *see* Registration of
engineers
selection of, 111
to societal problems, consciousness
of, 13
solicitation of, 110
supplanting of
engineering societies' prohibitions against, 58
example of, 45, 64
unethical conduct of, 52–66
United States Army Corps of, 76
Engineers Council for Professional
Development (ECPD), 10–11,
36
Environmental impact, 63–64, 116
as affecting property evaluation, 192
Environmental studies, 28
Equal opportunity regulations, 128
Equity jurisprudence, 87
Estoppel
agency by, 119
promissory, 103–104
Ethical conduct laws, federal enforcement of, 29
Ethical standards in a democratic
society, 7
Ethics, 1–9
absolute *versus* relative, 2–3
in business, 23–27
codes of, ancient
Buddha, teachings of, 8
the Code of Hammurabi, 8
Confucianism, 8–9
Ten Commandments, 6–7

divine law, compliance with, as
theory of, 2
early American examples of, 34
enforcement of, 2, 29
for engineering societies, 30–33
in government, 27–31
human rights, as related to, 4
legal, 19–23
medical, 17–18
professionalism, as related to, 1, 4–5,
10, 31
references, 9
self-discipline, as related to, 4
training and education in, 52–53
Executed contracts, 99

Feasibility reports, 63, 69, 111
Federal antitrust suits, as affected by
interstate commerce, 46
Federal Bureau of Reclamation, 76–77
Federal Circuit Court of Appeals, 90
Federal court system, 89–90
Federal highway system, 76
Federal judges, 89
Fixed-fee method for arriving at
engineering fees, 112
Fixed-percentage method for arriving
at engineering fees, 111
Foakes rule, as affecting contract law,
103

General contractors, 120, 127, 124–134,
135, 177
Government
ethics in, 27–31, 64–65
regulation of business, 24–25
Government contracts
bidding, public announcement of
invitation to, as requirement
of, 128
brand names, restrictions against
use of, 141–142
Government employees, 28
bribery of, 28, 55, 65, 117

Hammurabi, the Code of, as ancient
code of ethics, 8